AMPHIBIAN AND REPTILE ADAPTATIONS TO THE ENVIRONMENT

Interplay Between Physiology and Behavior

AMPHIBIAN AND REPTILE ADAPTATIONS TO THE ENVIRONMENT

Interplay Between Physiology and Behavior

edited by

Denis Vieira de Andrade
Instituto de Biociências, UNESP - Univ Estadual Paulista,
Departamento de Zoologia, Rio Claro, SP, Brazil

Catherine R. Bevier
Department of Biology, Colby College, Waterville, ME, USA

José Eduardo de Carvalho
Federal University of São Paulo, Brazil

CRC Press
Taylor & Francis Group
Boca Raton London New York

CRC Press is an imprint of the
Taylor & Francis Group, an **informa** business

CRC Press
Taylor & Francis Group
6000 Broken Sound Parkway NW, Suite 300
Boca Raton, FL 33487-2742

First issued in paperback 2020

© 2016 by Taylor & Francis Group, LLC
CRC Press is an imprint of Taylor & Francis Group, an Informa business

No claim to original U.S. Government works

ISBN 13: 978-0-367-57475-8 (pbk)
ISBN 13: 978-1-4822-2204-3 (hbk)

Visit the Taylor & Francis Web site at
http://www.taylorandfrancis.com

and the CRC Press Web site at
http://www.crcpress.com

Contents

Preface

The first words about this book were spoken soon after the closing of the symposium we organized as part of the scientific program of the 7th World Congress of Herpetology held in August 2012 in beautiful Vancouver, British Columbia. While still waiting to leave the presentation room, the speakers and symposium organizers gathered, as usual, to prattle about future collaborative research projects and plans that would surely happen in the short term. One of these plans indeed has come to fruition in the form of this book, composed primarily of many of the presentations delivered at that session. As such, this book assembles a diversity of topics related to the theme of the interplay between behavior and physiology as amphibians and reptiles interact with the environment. Expert contributors have worked generously to provide integrative and comprehensive reviews of subjects they have spent many years studying. The resulting chapters will hopefully promote new insights for students first learning about these topics without compromising the rigor and depth expected by experts. The chapters in this book can be read in any order, and we appreciate that not all titles will appear equally appealing to everyone at first glance. However, just as happens in scientific meetings, in which we sit through a whole session waiting for a particular presentation, our interest is often captured by different talks. We have no doubt that, if you are holding this book, you will find something relevant in every chapter. We want to thank Anthony Herrel for mentioning our inconsequential murmurs about this book to John Sulzycki, Senior Editor, Taylor & Francis, and to John for believing in us. As editors, we want to express our utmost appreciation to all the contributing authors of this book, especially given the time dedicated to preparing their respective chapters, as well as to the funding agencies that have supported our research activities over the years. Finally, we thank our mentors and families for their support and dedicate this book to them.

Denis, Cathy, and Zé
Rio Claro, October 31, 2015

Editors

Denis Vieira de Andrade earned his undergraduate degree in biology in 1992. He earned MS and PhD degrees in zoology in 1995 and 1998, respectively, both at the University of São Paulo State in Rio Claro, São Paulo, Brazil. Since 1997, he has worked at the same institution where he is now an associate professor of animal physiology in the Department of Zoology. His research interests focus on the ecophysiology, natural history, behavior, and conservation of amphibians and reptiles.

Catherine R. Bevier earned her BS in biology at Indiana University and PhD in ecology at the University of Connecticut. She is an associate professor of biology and has been at Colby College in Waterville, Maine since 1999. Bevier's current research focuses on behavioral and physiological ecology of amphibians and investigations of the complex relationship between frogs and the pathogenic chytrid fungus.

José Eduardo de Carvalho graduated with a BS in biological sciences at the University of São Paulo, Brazil, where he also earned his MS and PhD in animal physiology. He was a postdoctoral fellow at São Paulo State University in Rio Claro, Brazil and at the University of British Columbia in Vancouver, Canada. He is currently a professor of comparative animal physiology at the Federal University of São Paulo (UNIFESP), Campus Diadema, conducting research in ecophysiology and comparative biochemistry.

Contributors

Jens Frederik Dahlerup
Department of Hepatology
and Gastroenterology
Aarhus University Hospital
Aarhus C, Denmark

Eleonora Aguiar De Domenico
Departamento de Fisiologia
Universidade de São Paulo
São Paulo, Brazil

Sanne Enok
Department of Bioscience
Aarhus University
Aarhus C, Denmark

Peter Funch
Genetics, Ecology and
Evolution
Department of Bioscience
Aarhus University
Aarhus C, Denmark

Fernando R. Gomes
Departamento de Fisiologia
Universidade de São Paulo
São Paulo, Brazil

Aksel Kruse
Department of Surgery
Aarhus University Hospital
Aarhus C, Denmark

Harvey B. Lillywhite
Department of Biology
University of Florida
Gainesville, Florida

Alexander G. Little
Donnelly Centre for Cellular
and Biomolecular Research
The University of Toronto
Toronto, Ontario, Canada

William K. Milsom
Department of Zoology
University of British Columbia
Vancouver, British Columbia,
Canada

Edward J. Narayan
School of Animal and Veterinary
Sciences
Charles Sturt University
Wagga Wagga, NSW, Australia

Carlos A. Navas
Departamento de Fisiologia
Universidade de São Paulo
São Paulo, Brazil

Frank Seebacher
School of Life and Environmental
Sciences
The University of Sydney
Sydney, NSW, Australia

Lasse Stærdal Simonsen
Department of Bioscience
Aarhus University
Aarhus C, Denmark

Tobias Wang
Department of Bioscience
Aarhus University
Aarhus C, Denmark

chapter one

Behavior and physiology
An ecological and evolutionary viewpoint on the energy and water relations of ectothermic amphibians and reptiles

Harvey B. Lillywhite

Contents

Imagine one has the rare fortune of watching a cheetah stalking its prey—say an antelope—and then, instantly, breaking into an "all-out" run, which over a short course brings down the antelope in a spectacular life-or-death capture. What one witnesses is a quintessential expression of impressively well-orchestrated physiology. Adaptive evolutionary processes have resulted in a well-integrated suite of functions involving cardiorespiratory transport, metabolic energy production, and nervous coordination of visual and tactile stimuli resulting in rapid and precise muscle activations producing the spectacular behavior.

What animals "do" has been observed for centuries, and of course was central to much of the work of Charles Darwin and other early and

1

influential naturalists. As the scientific study of behavior grew into a more modern and defined discipline (ethology) during the last century, interest arose in the mechanisms of behavior, and thereby physiological investigations contributed importantly to the understanding of behaviors. Consideration of behaviors as phenotypic traits with genetic components of variation invited the application of phylogenetic methods in evolutionary studies addressing the role of natural selection in producing adaptation and reaction norms of such traits. Of course, genes do not themselves code for particular behavioral traits. Rather, they determine the proteins, expression, and regulation of the structural and physiological features that produce the behaviors we observe. And—as is well known to readers of this book—behavior has a very important influence on all aspects of physiology, so the two disciplines are tightly coupled. Patterns of neural activity can be mapped onto behavior, and these same neurons can be genetically tagged. A combination of technologies that are used in such mapping can reveal which neurons constitute circuits for specific behaviors (O'Leary and Marder 2014).

While a general thesis of this chapter is that physiology is the important underpinning of all behavior, readers will appreciate that the behaviors of ectothermic vertebrates are especially sensitive to environmental influences on physiology. A century ago, it was generally assumed that body temperatures of amphibians and reptiles passively followed that of the environment and that thermally associated physiology and behavior were obligatorily both ectothermic and poikilothermic (Pough 1974). This viewpoint subsequently changed, and the current understanding of thermal biology is multidisciplinary, integrative, and includes a paradigm that recognizes thermal sensitivity of virtually all aspects of physiology and behavior (Huey and Stevenson 1979; Bennett 1980; Huey 1982; Angilletta et al. 2002; Hochachka and Somero 2002).

Thermal physiology and behavior: A historical perspective

The historical pathway to understanding the complexity of thermal biology had origins that were based largely in observations of behavior. Early writings in the previous century established that temperature limited the latitudinal distribution of reptiles (e.g., Pearse 1931), while Klauber (1939) further recognized that temperature limited the *activity* of reptiles. His influence suggested that by actively seeking or avoiding particular thermal environments (earlier demonstrated for lizards in the laboratory by Weese [1919]), temperature (or heat) could be regarded as an important resource for ectotherms such as desert reptiles (see Turner 1984). Walter Mosauer and Raymond Cowles were both influenced by Lawrence

Klauber. Mosauer, an emigrant from Austria with experience and interests in deserts, spent time with both Raymond Cowles and Charles Bogert in California deserts where they principally investigated the locomotion of snakes. Based on observations that snakes suffered and died quickly when forced to remain in the desert sunlight, Mosauer became interested in temperature and published research demonstrating that both snakes and lizards living in deserts were not especially tolerant of high body temperatures (Mosauer and Lazier 1933; Mosauer 1936). Subsequent research by Raymond Cowles reaffirmed the findings of Mosauer and further established that squamate reptiles have specific preferences for body temperature (Cowles 1939; Cowles and Bogert 1944) (Figure 1.1).

This brief early history of the study of behavioral thermoregulation in reptiles illustrates how observations of behavior in the natural settings where animals actually live advances the conceptual understanding of hypotheses related to a central interplay between behavior and physiology. Similar observations were made by various field biologists who investigated amphibians (Brattstrom 1963). Subsequent research representing outgrowths from the earlier work took numerous directions having broader applications in biological thought, theory, and their collective influence. Importantly, much work related to the influence of temperature on ectotherms produced some major paradigms that became influential in evolutionary studies (Feder et al. 2000; Angilletta et al. 2002). Some of the more significant and enduring investigative spin-offs included: (1) identification of numerous behaviors whereby ectotherms avoid, seek, or alter the influence of the physical environment on body temperature

Figure 1.1 Common collared lizard, *Crotaphytus collaris*, basking on a rock in southeastern Utah. This lizard was basking, thermoregulating, and active during mid-morning hours during late spring. (Photograph by H.B. Lillywhite.)

(see Huey 1982; Lillywhite 1987; Hutchison and Dupre 1992); (2) understanding how the physiological influence of body temperature has important interplays with behavior (Van Mierop and Barnard 1976, 1978; Bartholomew 1982; Harlow and Grigg 1984; Turner 1987; Hutchison and Dupre 1992); (3) demonstrations that body temperature influences organismal performance (Lillywhite et al. 1973; Bennett 1980; Angilletta et al. 2002) (Figure 1.2); (4) appreciation from theory that thermoregulation has costs and benefits (Huey and Slatkin 1976); (5) understanding how thermal physiology and behavioral traits can be plastic (Kingsolver and Huey 1998) (Figure 1.2); (6) linking the roles of genetics and phylogeny to physiological and behavioral traits (Garland 1988; Garland et al. 1991; Garland and Carter 1994); (7) integrated modeling of energy balance (biophysical modeling) for animals and their environment (Porter and Gates 1969; Porter et al. 1973; Tracy 1976; Kearney and Porter 2009); and (8) developing applications of physiology and behavior to conservation and predicting the impacts of climatic change (Mitchell et al. 2008; Huey et al. 2012; Cooke et al. 2013, 2014; Lillywhite 2013).

Thermally sensitive functions in both physiology and behavior can exhibit lability, or shifts in reaction norms, related to environment (e.g., acclimatization; Figure 1.2; see Chapter 2), or can evolve over longer time

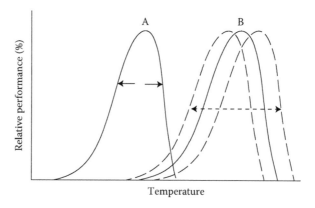

Figure 1.2 Theoretical, exemplary performance curves for two hypothetical species (A and B) adapted to living in different thermal environments. The horizontal arrows indicate the arbitrary breadth of each performance curve, which depicts relative changes in the chosen "performance" variable (e.g., sprint speed) as a function of temperature. The shifts of function represented by dashed curves in species B depict hypothetical changes of response attributable to thermal acclimatization (or acclimation). The dashed horizontal arrows representing species B reflect the total increase in performance breadth that is attributable to thermal acclimation. (Lillywhite, H.B., *How Snakes Work: Structure, Function and Behavior of the World's Snakes*, Figure 4.14, 2014b, by permission of Oxford University Press.)

scales. Changes within an individual can involve *de novo* mutations, coding and regulatory changes in standing genetic variation, masking or unmasking of allozymes, alterations of cell membranes and the internal cellular environment, changes in genetic expression, angiogenesis, and many other phenomena (Somero et al. 1996; Hochachka and Somero 2002; Jones et al. 2012). The evolutionary pathways by which such changes occur are variable, but there is clear potential for physiology and behavior to evolve rapidly when subjected to selection (Lukoschek and Keogh 2006; Rosenblum 2006; Edgell et al. 2009; Garland and Rose 2009; Rohner et al. 2013; Sanders et al. 2013). Thus, in a world that changes because of geological processes, global climatic anomalies and warming trends, emerging pathogens, transoceanic exchanges of biota (invasive species problems), and numerous other anthropogenic impacts (Rohde 2013), physiologists and ethologists should increasingly view their work and their investigative principles from a perspective that appreciates a biological fabric and ecological landscape that are fluid (e.g., Rosenblum 2006; Mitcheletti et al. 2012). This viewpoint is different from the attitudes of earlier biologists who, just a few decades ago, worshipped "mean" values and regarded them largely as unchanging attributes and de facto descriptors of taxa (Bennett 1987).

Energy, water, and behavior

Energy and water are fundamental requirements of living organisms. Moreover, variation in the temporal availability of energy and water can determine the functional properties and physiological performance of individuals. Considering both evolutionary and ecological scales, the availability of energy and water influences nutritional status, body condition, demography, reproduction, and ultimately fitness of animals across variable landscapes. Fundamentally, any system—whether it be an individual organism or a population of a species—cannot remain in a steady state and will "run down" if energy becomes limiting or absent. The same argument applies to considerations about water, with exceptions of unusual instances of anhydrobiosis observed in certain invertebrates (Alpert 2006). While transient states often characterize organisms at any point in time, assumptions about "adaptation" often assume that (1) energy is limiting and (2) adaptive characters are, effectively, in a condition of a steady state. Yet, we must appreciate that there are many circumstances in which energy is not limiting and the numbers and conditions of organisms are determined by other factors such as social behavior (Stewart and Pough 1983) (for mammals, see also Armitage 2014), disease (Pounds et al. 2006; Blaustein and Bancroft 2007), or availability of other resources, for example, water and breeding sites (Stewart and Pough 1983; Donnelly 1989). Of course, behavioral interactions among individuals (competition, partitioning of resources, and the like) can interplay with energy requirements as they relate to the "compressibility" of a species niche (Schoener 1974).

Energy, ectothermy, and behavior

Insofar as energy is the fundamental currency of "living," its acquisition and expenditure underpins the entire behavioral repertoire of complex animals, including amphibians and reptiles (see also Chapter 6). Scientific interest and investigation of this fundamental principle emerges historically in at least two important ways. First, the demands for energy are important drivers in the evolution of demographic parameters that have established the complex associations of populations and species within communities and various environments (Pough 1983; Beaupre 2002; McNab 2002; O'Connor et al. 2006). Second, quantitative studies have described the energetic costs associated with specific behaviors such as those involved in reproduction, locomotion, feeding, defense, etc. (e.g., Secor and Nagy 1994; Secor 1995). Hence, the abundance, diversity, and social interplay (or not) of animals have important determinants based in energy and behavior. Elton (1927) argued that understanding the distribution and numbers of animals in nature requires an appreciation of "what animals do" and both "the circumstances" and "limiting factors" related to this.

Animals must maintain a balanced energy budget over time, and for ectothermic amphibians and reptiles the challenges related to acquisition of energy are clearly expressed in alterations of behavior. Well known, but instructive examples of behavioral responses to energy balance, are the behaviors of ectotherms in response to environments with a marked seasonality in physical conditions and availability of food (Figure 1.3). With the exception of some aquatic species (e.g., Gregory 1982; Ultsch 1989), the impact of seasonally harsh conditions generally curtails the activity and seasonal energetics of ectotherms. The lizard *Uta stansburiana*, for example, exhibits higher field expenditures of energy during spring and early summer but lower expenditures in winter (Nagy 1983). Such differences in the seasonal expenditure of energy reflect behavioral inactivity, digestive limitations, and lesser availability of food during winter than during summer. Rates of food intake during winter can be dramatically curtailed (about 5% of that in summer in the scincid lizard *Lampropholis quichenoti*: Gallagher et al. 1983) or absent altogether during conditions of brumation. In addition, decreased energy expenditure during field conditions may occur in relation to drought or related limitations in the availability of prey (e.g., Nagy and Shoemaker 1975; Carvalho et al. 2010; and see below). The physiological underpinnings of the behavioral changes impacted by energy resources and expenditure involve a suite of interactions related in large part to thermal constraints on the neural

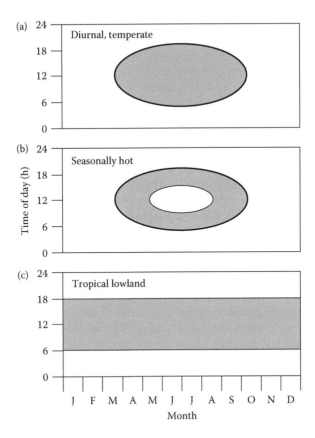

Figure 1.3 Seasonal variation in the potential time available for activity (shaded areas) of diurnal lizards, as constrained by thermal physiology and the thermal environment in the northern hemisphere. (a) Elliptical activity characteristic of numerous diurnal lizards living in the temperate zone. (b) Activity characteristic of lizards living in seasonally hot environments, such as deserts, where high summer temperatures inhibit activity during midday. (c) Relatively invariant daily activity characteristic of some tropical lowland lizards. (Revised and redrawn based on Figure 1.2, Adolph, S.C. and W.P. Porter. 1993. *Am. Nat.* 142:273–295.)

activation and function of muscle, thermal constraints on digestion, circannual changes in endocrines, timing of life cycle events, and other factors that are not well studied.

In contrast to cold-temperate environments, ectotherms in tropical and warm-temperate environments remain largely active throughout the year as long as food and water are available (Figure 1.3). However, tropical environments are generally more seasonal than might be indicated by judgments based on air temperatures. Thus, seasonality of activity and energetics are found in varanid lizards living in tropical Australia despite the persistence of high daytime temperatures throughout the year (Christian et al. 1995). Pronounced seasonal variation of activity and expenditure of energy is particularly well documented with respect to seasonal drought when both water and prey become lacking. Generally, terrestrial amphibians and reptiles seek shelter (such as burrows) and reduce expenditure of energy when water is not available (Christian et al. 1995; Seebacher and Alford 1999; McArthur 2007). The extent of seclusion, however, depends on the local conditions of the habitat. The varanid lizard *Varanus panoptes* is intensely active during the dry season when it walks long distances during foraging along the receding edges of floodplains. However, this species becomes inactive when the floodplain dries completely or if they are living in woodlands that are considerable distances removed from the receding water edges (Christian et al. 1995). Generally, the seclusion of both amphibians and reptiles in response to seasonal or unusual drought is a well-known phenomenon, although the details of physiological "triggering" in such changes of behavior are in need of much further investigation.

The behavioral and physiological "management" of energy resources is complex and dependent on the conditions of the environment as well as the internal state of an animal (Beaupre 2002; Milsom et al. 2012). Nearly all vertebrates reduce body mass during prolonged fasting, but different organs and body compartments may lose mass at different rates and are probably prioritized (McCue et al. 2012). Starving snakes may increase or decrease energy expenditure depending on the environment, behavioral activity, and related tradeoffs. Snakes that are subjected to starvation in the laboratory exhibit significant metabolic suppression, even when measured rates of metabolism are corrected for concomitant reductions of body mass (McCue 2010). The mechanisms involved at a molecular level remain largely unexplored. Like many animals, snakes oxidize endogenous lipids during starvation. However, ketone bodies do not consistently increase, and levels of glucose do not consistently decrease, in blood (McCue et al. 2012). Endogenous protein is also oxidized as fuel but at much lower rates than carbohydrates or lipids. Starving snakes continually produce uric acid, which is voided as discrete events. Interestingly, in many species of snakes, the feeding rates of females decrease during reproduction, and gravid females may become anorexic (Madsen and Shine 1993). Feeding rates decrease progressively with increasing follicle size in reproductive female sea kraits, and feeding ceases altogether as the eggs develop. The pattern of a progressive reduction of feeding with egg development suggests that a "threshold" effect is lacking and complements the hypothesis that bodily distension impedes locomotor effectiveness during foraging and increases vulnerability to aquatic predators (Brischoux et al. 2010).

When conditions related to prey availability, water, and habitat are appropriate, behavioral thermoregulation favors optimal cellular and enzymatic function as well as behaviors such as locomotor performance (Figure 1.2). *Ameiva* lizards in the tropical West Indies thermoregulate effectively within variable thermal habitats throughout their daily activity, which tends to peak during the morning (Gifford et al. 2012). Many temperate, subtropical, and tropical lizards associated with relatively open habitats spend time basking and maintain relatively high body temperatures that are generally above the prevailing ambient conform temperatures (Ruibal 1961; Porter et al. 1973; Huey 1982; Vitt et al. 1998). However, other tropical lizards live within deeply shaded forests where behavioral thermoregulation is more difficult and body temperatures conform more to prevailing air temperatures (Inger 1959; Ruibal 1961; Huey 1982; Hertz 1992). Indeed, nonbasking species constitute a major component of the diversity of lizards in the neotropics (Huey et al. 2009). Similar trends probably characterize amphibians and snakes, although data for these groups are less well analyzed. The complexity of factors that are influential with respect to motivation for various behaviors is not well understood, especially in reclusive or elusive species of ectotherms. Tendencies of lizards to exhibit basking behavior are generally associated with coadapted traits having thermal sensitivity, show phylogenetic conservatism, and also correlate with habitat (Angilletta et al. 2006; Huey et al. 2009). The benefits of thermoregulation are not well understood in tropical reptiles, which require further investigation especially with respect to tradeoffs of costs and benefits and ecological circumstances (Luiselli and Akani 2002; Bovo et al. 2012).

Climatic warming

Reptiles provide an instructive assemblage of taxa for assessing and predicting the impact of climatic changes owing to the physiological and behavioral features that are characteristic of this group (Lillywhite 2013). Here, it merits some comments related to global warming insofar as 2/3 of the global variation in species richness of reptiles can be explained, in some analyses, by temperature alone (Qian 2010). Increasing temperature will have complex and multiple effects on reptiles even if considered the sole driver in climatic change. Past and future climatic changes are, or will be, represented by extreme events, gradual long-term changes, and local anomalies. Both the nature and magnitude of warming impacts will affect physiology both directly and indirectly as modified by plasticity of genetic, cellular, and organismal system response and, importantly, by behavior. Although past climatic disruptions and attendant ecological changes have been severe, the effects on organisms varied greatly among taxa because of differing requirements and adaptability. Acclimation or physiological plasticity increases resilience of ectothermic animals to climatic change (Seebacher et al. 2015; Chapter 2), whereas the efficacy of genetic adaptation depends importantly on the interplay between generation time and the rate of climatic change (Lande 2009; Hoffmann and Sgrò 2011). Temperature fluctuations are predicted to increase with climatic changes, and these may not provide an adequate driver of directional selection.

Survival and adaptation to climatic changes are likely to be favored by small body size, low metabolic requirements, and behavioral plasticity in a diversity of microenvironments and landscapes (Lillywhite 2013). Underground shelters provide important thermal as well as hydric buffers and can be vital resources affecting the survival and use of habitats by ectothermic vertebrates (Bruton et al. 2014). As the climatic and biotic features of landscapes change, limitations or lack of shelters can perturb behaviors and increase levels of stress hormones with cascading effects on physiological performance (Bonnett et al. 2013). Thus, environmental changes can produce subtle and interactive consequences important to understanding behavioral changes and other processes related to fitness (Moore and Jessop 2003; see Chapters 7 and 8).

Data from reptilian taxa generally support the hypothesis that changes in global temperature largely affect biodiversity at higher latitudes and altitudes, although this is possibly a premature or incorrect generalization (Lillywhite 2013). Moreover, because of changes to habitat and complexity of anthropogenic factors, it is difficult to attribute causation to climate as being distinct from a myriad of other effects. Nonetheless, increasing numbers of studies suggest that numerous declines and extinctions of reptilian taxa link causally with climatic change (Raxworthy et al. 2008; Reading et al. 2010; Sinervo et al. 2010; Lillywhite 2013). Extinctions of species, declines of populations, and biogeographical changes might be comparatively high in the tropics because of high species richness and incidence of endemism. Extinctions of high-altitude viviparous lizards in Mexico are expected to exceed those of more lowland species of lizards (Sinervo et al. 2010).

Poleward and altitudinal migration has been a focus of some attention and appears to be an important biogeographical factor that is driven by past and future changes of climate, particularly in temperate regions and in tropical mountains (Bush 2002; Hickling et al. 2006; Rull and Vegas-Vilarrúbia 2006; Raxworthy et al. 2008; Chen et al. 2011). Details concerning the behavioral responses of animals and their interplay with the expenditure and acquisition of energy (related to climatic changes) are largely unknown. Do latitudinal and altitudinal shifts in distribution occur at rates determined by the usual dispersal and behaviors of animals, or do some activities become more intense in response to changing climate? Also, reptiles inhabiting cool environments might actually benefit from increased warming with respect to advantages for extended activity time and energy acquisition (Chamaillé-Jammes et al. 2006; Kearney et al. 2009). High-altitude *Trimeresurus gracilis* in Taiwan are predicted to have energetic advantages associated with climatic change, and biophysical modeling of montane niches demonstrates an increased digestive capacity related to extended time of activity and to energetic benefits arising from the warming of habitats (Huang et al. 2013). Positive changes in body size, female fecundity, and survival rates have been documented for the high-altitude populations of the European lizard *Lacerta vivipara* experiencing longer active seasons related to recent climatic warming (Chamaillé-Jammes et al. 2006). On the other hand, upslope movement of forest cover could create disadvantages for some heliothermic species of reptiles, and thus the impact of warmer climates will be greatly affected by future patterns of vegetation (Huang et al. 2014). More generally, the impact of climatic warming on thermoregulating reptiles will depend on how changes in vegetation affect both shade and basking sites and whether animals can alter the seasonal timing of activity and reproduction (Kearney et al. 2009).

In contrast with high-altitude reptiles, species living in tropical lowlands might be threatened by decreased activity time and by impairment of physiological functions as temperatures increase (Tewksbury et al. 2008; Huey et al. 2009; Sinervo et al. 2010). Several studies have raised concern for the vulnerability of tropical ectotherms, and declines in populations of lowland forest species have been documented (Huey et al. 2009). The difficulty in predicting responses to climatic warming, however, is to accurately forecast how habitats will change and to what extent such changes will affect biotic interactions (Mitchell et al. 2008). The impacts of climatic change will comprise complex and myriad interactions, including physiological stress, altered productivity and food web dynamics, shifts in species distributions, behavioral interactions, alterations of overlap affecting activity and acquisition of energy, increasing incidence of disease, and the interplay between genetic changes and plasticity of expression (Lillywhite 2013; Chapters 7 and 8). Because environmental temperatures influence life history phenomena, global warming could have profound effects on reptilian ecology and evolution (Meiri et al. 2013).

Water and behavior: Amphibians

Although generally given less attention than energy, water also is a very important driver of behaviors related to the acquisition and uses of energy, activity, and ultimately survival. The hydration status of amphibians and reptiles can vary greatly, and departures from steady states of water balance profoundly influence what these animals do—when, where, and with what degree of physiological performance in many different contexts (Hillman 1982, 1984; Gatten 1987; Preest and Pough 2003; Titon et al. 2010). Amphibians and reptiles have become adapted to harsh environments, and various species exhibit extreme changes in behavior related to variation in the availability of water.

Inactivity or torpor characterizes many species of amphibians (also various reptiles) that estivate to avoid the harshest conditions of water limitations in semiarid or arid environments (Carvalho et al. 2010). Rates of energy expenditure (reflected in measurements of rates of metabolism) in estivating amphibians can depress to less than 30% of measured rates in resting individuals (Guppy and Withers 1999; Carvalho et al. 2010), and some frogs can theoretically remain in such a dormant state for several years (van Beurden 1980). In the frog *Cyclorana alboguttata*, whole animal metabolism is reduced by 82% within 5 weeks of estivation (Kayes et al. 2009). Seymour (1973) estimated that spadefoot toads (*Scaphiopus couchii*) could survive at least 2 years of drought using fat stores alone. Metabolic depression is indeed important insofar as these amphibians need to synchronize emergence and reproduction with unpredictable rain patterns and occasionally prolonged droughts. Reproduction and metamorphosis must be completed in transient ponds of water (Bentley 1966). Successful reproduction requires that females have viable eggs and males have viable sperm as well as adequate stores of energy to fuel intense calling to attract females, amplexus, and in some species, use of the legs to build a foam nest (see Chapter 3). Thus, energy stores must be managed to last over the period of estivation and, from time to time, droughts lasting more than 1 year. Strategies of behavior no doubt evolved early in the evolutionary histories of several groups to enable underground estivation in appropriately selected locations, adequate digging capabilities, successful emergence and location of temporary pools, and overcoming the physiological challenges of underground dormancy such as drying soils and hypoxia. However, understanding how reproduction and timing of emergence coevolve in unpredictable environments remains a subject that requires much further investigation (Carvalho et al. 2010).

Physiological adaptations that complement behavior in relation to estivation include downregulation of molecular and cellular physiological function, regulation of energy stores, upregulation of protein catabolism and synthetic pathways important for formation of urea as an osmolyte, lack of skeletal muscle atrophy, and formation of a cocoon (Carvalho et al. 2010) (Figure 1.4). Despite extended periods of inactivity and fasting, dormant frogs experience very little atrophy of muscle related to a complex suite of biochemical changes. Recent studies have demonstrated that complex changes of gene expression involving transcriptional regulation of genes are associated with cytoskeletal remodeling, avoidance of oxidative stress, energy metabolism, and apoptotic signaling (Reilly et al. 2013). Thus, muscle function in estivating frogs (*C. alboguttata*) results from expression of genes in several major cellular pathways that are critical to survival and viability of cells.

Amphibians exhibit many features that reflect evolutionary transitions between aquatic and terrestrial environments. Adaptations of

Figure 1.4 (a) Intact cocoon with head portion peeled back on an estivating *Cyclorana australis*. This frog had been induced to form a cocoon for several weeks in the laboratory. The fore part of the cocoon has been peeled back to stage the photo. Normally, the cocoon would soften from the bottom when exposed to water, and the frog would work itself out of the cocoon much like a shed skin. (b) Isolated cocoon after having been removed from an estivating *Cyclorana australis*. (Photographs courtesy of Christopher Tracy (a) and Scott Reynolds (b).)

behavior in this context are reflected in both evolutionary history and specializations of extant forms. Complex life histories with aquatic larval stages and a subsequent transition to the terrestrial characters of adults involve profound changes of behavior that parallel developmental changes in morphology and physiology and the respective adjustments to life in very different habitats. Here, I will comment further on the skin as a key feature of amphibians and a principal determinant of their behavior.

In contrast with most reptiles, amphibians have integument that is relatively thin and pliant, lacking extensive keratin, and having permeability properties that allow significant gaseous exchange with the environment. Many species of amphibians possess integument that evaporates similarly to a film of free water, although such a property may not represent the majority of more than 7300 species (Lillywhite 2006; Tracy et al. 2010). In species where water evaporates relatively freely across the skin, behaviors that enable the acquisition of water or reduce losses of water are important. These involve nocturnal activity, posturing behaviors, and careful selection of microhabitat.

Some amphibians maintain almost permanent contact with water, and they enter a state of estivation that includes formation of cocoons when water sources dry out (Ruibal and Hillman 1981; Withers 1995, 1998) (Figure 1.4). Other species have periodic contact with water. Anuran amphibians "drink" cutaneously by absorbing water across a "seat patch" or specialized region of morphologically and physiologically distinct ventral pelvic skin (McClanahan and Baldwin 1969; Bentley and Yorio 1979; Brekke et al. 1991). Cutaneous drinking elicits specialized behaviors, such as the manner of walking over dry or wet substrates (Stille 1951), and a "water absorption response" that maximizes the area of ventral skin contacting water (Still 1958; Brekke et al. 1991). The absorption of water from moist soils may involve the transfer of water by capillarity from a matrix of soil particles, with possible enhancement of transfer attributable to epidermal sculpturing (Lillywhite and Licht 1974; Hillyard 1976; Brekke et al. 1991). Angiotensin II stimulates cutaneous drinking in terrestrial toads, and the mechanism that regulates hydration in amphibians appears to be homologous to thirst mechanisms that are known in other vertebrates. Thus, physiological elements of the water absorption response could represent an important step in the evolution of thirst in terrestrial vertebrates (Hoff and Hillyard 1991; Tran et al. 1992).

A variety of anuran behaviors serve either to limit evaporative water losses from the skin or to maintain the skin in a wetted condition using water from external sources (e.g., Lillywhite and Licht 1974). Behaviors that are expressly related to this fundamental requirement reflect the evolutionary history of the group, aquatic larval stages in many species, and retention of skin that is characteristically porous to water because of limited keratinization and lack of a lipid–keratin barrier that is present in amniotes. Keratin was presumably present in basal amphibians (Maderson 1972), but neither extensive keratinization nor synthesis of β-type keratins are characteristic of extant amphibians. The limited keratinization of the epidermis of amphibians is curious but incurs a critical constraint, possibly attributable to developmental canalization during aquatic life stages (Lillywhite and Mittal 1999). Such condition of the skin might be more related to a lack of genetic expression than to gene deficiencies, insofar as specialized structures such as cornified protrusions (spade-shaped) reflect an inherent capacity for extensive synthesis of keratin. Nonetheless, it seems that plesiomorphy in relation to integumentary morphology is an important driver of amphibian behavior. These properties of the integument strongly constrain both the activities and the distribution of amphibians (Lillywhite and Mittal 1999).

Insofar as amphibians are lacking the dense lipid–keratin complex that characterizes the stratum corneum of terrestrial amniotes (Lillywhite 2006), evolutionary responses to dry environments have produced similar protective effects attributable to water barriers having time-limited integrity external to the skin. Two principal means of "waterproofing" have evolved: (1) production of a temporary cocoon (Figure 1.4) and (2) production of a waxy film that is wiped over the skin surfaces and requires periodic renewal (Figure 1.5). In both cases, the influence on behavior is

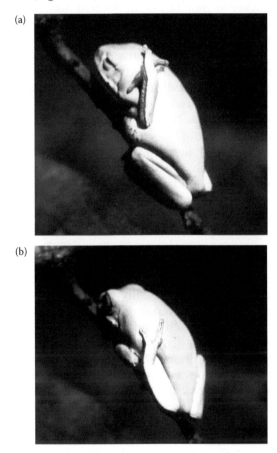

(a)

(b)

Figure 1.5 Wiping movements characteristic of orange-legged leaf frog, *Phyllomedusa hypochondrialis*, common to the dry Chaco region of South America. (a) Each forelimb wipes separately to spread secreted lipids over the dorsal forebody and head. (b) Each hindlimb wipes secreted lipids over the central and rear dorsum of the body. Other wiping movements that spread lipids over the ventral body and limbs are not shown. Once the lipid secretions are spread over the skin surfaces, rates of evaporative water loss are reduced to as little as 4% of that from a free water surface (see Gomez et al. 2006). (Photographs by H.B. Lillywhite.)

profound because the integrity of either external barrier depends on a cessation of movement and activity.

Characteristic "cocoons" of amphibians have evolved independently in terrestrial taxa representing several lineages of anurans (Hylidae, Hyperoliidae, Leptodactylidae, Myobatrachidae, Pelobatidae, and Ranidae) and some salamanders (Reno et al. 1972; Ruibal and Hillman 1981; Etheridge 1990; Withers 1995, 1998; Christian and Parry 1997). These structures consist of multiple layers of shed epidermal stratum corneum and become thicker as new layers are formed and gradually harden to encase the entire body except for openings of the external nares. Cocoons of various species increase resistance to water passage from 10- to several 100-fold, and greatly retard evaporative water losses to air and hydraulic losses to soil. However, studies of Australian frogs (*Cyclorana australis*) demonstrate that cocooned individuals do not exchange significant amounts of water when placed on semisolid agar-solute substrates across a range of water potentials. The cocoon appears to act as a physical barrier that breaks the continuity between frog and substrate, suggesting that it functions to prevent liquid water loss to drying clay and loam soils (Reynolds et al. 2010). The direction and nature of water exchange will depend on the soil type and its moisture content, so cocoons of various species might retard dehydration differently in various microhabitats.

The composition of both skin secretions and cocoon material of *C. australis* was shown to contain 5%–10% neutral lipids and 78%–85% proteinaceous material (Christian and Parry 1997). Thus, the cocoon structure of this and possibly other amphibian species forms a layered lipid–keratin complex that is similar in essential architecture to the water permeability barriers of amniotes (Lillywhite 2006).

The second means of "waterproofing" amphibian skin involves the secretion of lipids onto the skin surfaces followed by elaborate and stereotyped wiping behaviors that spread the lipids to create a film that covers the entire body surface (Figure 1.5). The elaborate wiping of lipids onto body surfaces was first described in phyllomedusine frogs, which exhibit very low rates of evaporative water loss (Blaylock et al. 1976). These lipids consist largely of wax esters within a complex mixture, including triglycerides, hydrocarbons, free fatty acids, and cholesterol, and these are wiped to form a layer that is about 0.2 μm and 50–100 molecules thick (McClanahan et al. 1978). This film of lipids provides a barrier to water movement and greatly increases the resistance of the skin to evaporative water loss, enabling some arboreal species to remain exposed to sunlight during hot, dry weather when body temperatures approaching 40°C are tolerated (Shoemaker et al. 1987). The savings of water attributable to the external film of lipids is further complemented by the excretion of uric acid in "waxing" species (Shoemaker et al. 1972; Shoemaker and Bickler 1979).

What is especially dramatic is the stereotyped wiping behavior (Figure 1.5) that has evolved in concert with the production of waterproofing lipids and their secretion onto skin surfaces. Both wiping behavior and secretion of lipids from cutaneous glands have evolved in several species of anurans, conferring variable skin resistances generally greater than the resistance in skin of frogs that are tied closely to water (Lillywhite 2006). Precursors to the behavior, and the origins of cutaneous lipids, might have evolved in a number of different contexts, while the adaptive coupling of the two appear to be associated with arboreal habits and living in arid or ephemerally arid habitats where there is potential for dehydration stress. The cutaneous lipids of various species are secreted from specialized lipid glands (McClanahan et al. 1978), from mucus glands (Lillywhite et al. 1997a,b), or from granular glands (Barbeau and Lillywhite 2005) in different taxa. Moreover, both the complexity of wiping and the resistance of external lipid films exhibit variation among species of hylid tree frogs in Florida (Barbeau and Lillywhite 2005).

The water barriers of amphibian skin differ from those of amniotes in fundamental ways. First, amphibians evidently do not produce lamellar granules that are characteristic of keratinizing epithelia of amniotes (Lillywhite 2006). Second, the corneous layers of the epidermis of amphibians are too sparse to provide an effective lipid–keratin complex, with the possible exception of cocoons which have not been thoroughly investigated in this context. Third, lipid permeability barriers of amphibians are structured externally to the epidermis, whereas lipids previously hypothesized to be protective within the integument are not effective in comparison with the externally layered films (Lillywhite 2006). Finally, external films that function as a water barrier do so transiently and limit the activity of animals during periods of torpor or rest. Movements of body parts disrupt the integrity of such films, and wiping behaviors of the arboreal species secreting lipids periodically renew such films following bouts of activity.

What seems clear in evolutionary terms is that supraintegumentary permeability barriers have evolved multiple times in association with dehydrating environments. Cocoons can be formed in terrestrial–fossorial species representing 9 of 75 families of amphibians, and lipid films attributable to cutaneous secretion and wiping are documented in arboreal species of tree frogs representing 3 of 55 families of anurans (Lillywhite 2006). Wiping behaviors that have a waterproofing function are specific to environmental context and may well occur in numerous species that have not yet been observed carefully to document the behavior in association with the secretion of lipids. Because of the pliant nature of amphibian integument and its limited keratinization, external lipid barriers seem to provide the more effective and practical means of waterproofing in amphibians, while evolutionary pathways to other solutions appear to be constrained (Lillywhite and Mittal 1999; Lillywhite 2006).

In the case of both cocoons and secreted supraepidermal lipid films, the subsequent barrier function depends on the immobility of the animal because movements will disturb the structural integrity of the barrier (Lillywhite 2006). Hence, such permeability barriers are transiently associated with periodic states of seclusion and torpor in terrestrial species, or periods of rest when arboreal species assume a water-conserving posture. The adoption of arboreal habits among anuran species appears to require either availability of water (e.g., in rainforests) or some degree of resistance to evaporative water loss in combination with the periodic use of water-conserving postures and/or the evolution of large body size (Tracy et al. 2010).

Studies have indicated that although there is a strong ecological component that influences the evolution of comparatively high skin resistance to water loss among anurans, there is also phylogenetic influence (Prates and Navas 2009). The ecological factors importantly include both climate and behavior—especially whether a species is terrestrial and fossorial or arboreal. Within the genus *Litoria*, for example, cutaneous resistance correlates with phylogenetic position, but the greater values are associated with arboreal species (Young et al. 2005). Also, frogs of the genus *Phyllomedusa* exhibit the higher values for skin resistance among hylid species, and the phyllomedusines are highly arboreal (Wygoda 1984; Lillywhite 2006). This statement requires qualification, however, considering the terrestrial species *Pternohyla fodiens* forms cocoons (when fossorial) with even higher resistance than the wiped wax films of other arboreal hylid species (Withers et al. 1982; Withers 1998).

With respect to a context of phylogeny and behavior, it is important to note that closely related species of anurans are able to live in contrasting environments without evolving greatly different skin resistance. For example, two species of the ranid frog *Platymantis* having similar and negligible skin resistance live in terrestrial and arboreal environments, respectively (Young et al. 2006). Moreover, species of the bufonid genus *Rhinella* from contrasting environments in Brazil exhibit very low skin resistance—typical of other members of Bufonidae that have been investigated—yet one species (*Rhinella granulosa*) has successfully colonized the semiarid environment of the Caatinga (Prates and Navas 2009). Similarly, the bufonid species *Anaxyrus cognatus* (previously *Bufo cognatus*) has skin resistance that is relatively low and characteristic of its family, yet also occurs in xeric regions of northern Mexico and southwestern USA (Withers et al. 1984; Wygoda 1984). Both *A. cognatus* and *R. granulosa* have slightly elevated resistance in comparison with bufonid species from more mesic habitats, but the values are nonetheless close to zero and are possibly constrained by phylogenetic association. These data suggest the importance of behavior as well as physiology in colonizing xeric environments, with bufonids perhaps relying largely on behavioral hydroregulation related to the uptake and conservation of water (Navas et al. 2007; Prates and Navas 2009). Carlos Navas et al. emphasize that *Rhinella* have long seasons of activity in the semiarid Caatinga, suggesting that these amphibians have exceptional abilities to detect and extract water from soil, and this be a key aspect of behavior in the survival of the species. The Fiji tree frog, *Platymantis vitiensis*, exploits arboreal environments, but the environment is humid and these frogs live near streams and rivers (Young et al. 2006). Thus, physiological and behavioral traits relevant to the physiological ecology of amphibians (as well as other vertebrates) include a diverse array of strategies for the maintenance of water balance, and these undoubtedly include interactive tradeoffs related to gas exchange, energetics, reproduction, defense, and thermoregulation (e.g., Toledo and Jared 1995; Lillywhite and Mittal 1999; Prates and Navas 2009; Tracy et al. 2010).

Water and behavior: Reptiles

In comparison with amphibians, reptiles represent taxa, which in terms of phylogenetic history have evolved traits that confer a greater degree of independence from free water. Particularly important are oviparity and viviparity without complex life cycles, increased capacity for salt excretion, generally thicker integument with a greater degree of cornification, and greater resistance of skin to water exchange attributable to a specialized lipid–keratin barrier. Thus, reptiles occupy harsher desert environments in larger numbers than do amphibians, and importantly, a significant number of species have adapted to marine environments, whereas no amphibian is fully marine.

Here, I will discuss the physiology and behavior of reptiles in two environmental contexts where these animals are challenged to maintain water balance: deserts and ocean. Both environments are lacking in free drinking water, but the constraints are different. Marine environments impose an additional demand on osmoregulation related to salt balance, and the evolutionary transitions from terrestrial or freshwater environments to marine environments are more "abrupt" than is the invasion of deserts via range expansions through associated ecotones. On the other hand, low-latitude deserts impose thermoregulatory challenges because of higher temperatures compared with oceans, although refugia typically are available to small terrestrial ectotherms. Secondarily, marine reptiles generally rely on air-breathing, and therefore dormancy is not an option for animals that need to visit the surface of the ocean periodically to exchange respiratory gases. These differences in features of habitat have a variety of implications related to behavioral as well as physiological responses to environmental conditions.

Species in arid environments

Reptiles living in arid environments need to balance the need for conserving water with requirements for activity, management of energy, reproduction, and other bodily requirements and processes. Limitations of evaporative water loss are largely attributable to an effective water permeability barrier in the stratum corneum of the epidermis (Lillywhite 2006) and reductions of respiratory water loss in some species (e.g., Snyder 1971; Nagy and Seely 1993; Gienger et al. 2014). Reptiles are unable to excrete urine that is hyperosmotic to body fluids, but some species have evolved extrarenal salt glands that enable ion excretion with minimal water losses (Peaker and Linzell 1975; Dantzler and Braun 1980; Hildebrandt 2001). The importance of salt glands varies among species as well as individuals depending on diet and the relative status of ionic and osmotic balance (e.g., Hazard et al. 2010; Babonis and Brischoux 2012). Other water-conserving mechanisms include the excretion of uric acid, reabsorption of water in the hindgut or cloaca, sometimes relatively high thermal tolerance, and behavioral mitigation of dehydration attributable to seclusion or dormancy (Bradshaw 1986; Guppy and Withers 1999; Lillywhite 2006).

Limitations of water also can profoundly influence embryonic development and the quality of offspring, and can drive embryonic death (Vleck 1991; Belinsky et al. 2004; Du 2004; Brown and Shine 2005). Hence, multiple adaptations have evolved with respect to selection of nesting sites, parental care of eggs to minimize water loss, and eggshell structure either to minimize water loss of hard-shelled eggs in a desiccating atmosphere or to favor uptake of water in flexible-shelled eggs in humid environments (Belinsky et al. 2004; Shine and Thompson 2006; Stahlschmidt and DeNardo 2010).

Some desert reptiles have comparatively low requirements for energy (Beaupre 1993; Nagy et al. 1993; Beck and Lowe 1994; Beck 1995), and these have behavioral and ecological consequences that are shared among ectotherms generally (Pough 1980). Snakes, for example, are successful inhabitants of deserts. Their ability to consume large meals, relatively high storage capacity for energy, and comparatively low rates of metabolism reduce the necessity for feeding frequently (see Chapter 4). Physiologically, lower rates of metabolism incur lower rates of respiratory water losses attributable to reduced lung ventilation, but whether these savings are significant is somewhat debatable. In behavioral contexts, low energy requirements reduce foraging demands and can be further reduced by spending time in relatively cool retreats and avoiding time on warmer ground surfaces that elevate temperature and also increase demands for energy otherwise spent in locomotion, postural adjustments, social interactions, defense, etc. Rates of water flux generally follow patterns for field metabolic rates, and thus reduced activity (e.g., in response to seasonal drought) decreases water flux compared to reptiles in the same environment during wetter periods or in arid environments generally (Christian et al. 2003).

Numerous factors influence the savings of water in relation to behavior. The more important ones include temperature and water vapor pressure of selected environments, postural adjustments, responses to weather, timing of movements, area of direct bodily contact with features of refugia, presence and load of ectoparasites, breathing patterns, disease, nutritional status, and the health of the skin. The behavior of an animal will vary depending on the combination of conditions it encounters, and surely this explains (in part) the variation of behaviors we observe and, at times, the "surprise" appearances or activities of individuals that are seemingly "out of character." Examples of the latter might include underground movement or emergence of desert toads in advance of approaching storms and rainfall (Ruibal et al. 1969; Dimmitt and Ruibal 1980), and spectacular eruptions of sea kraits that emerge from terrestrial refugia to drink from rock pools that are formed during rainstorms following periods of drought (Bonnet and Brischoux 2008). Although temperature might seem to limit reductions of metabolism in tropical reptiles compared with temperate ones (which, e.g., overwinter at very low temperatures), nonetheless tropical species can achieve substantial energy savings by means of behavior or a combination of behavior and metabolic depression (Christian et al. 2003).

The extent to which physiological and behavioral plasticity is a part of water-conserving responses depends on evolutionary history and the constraints of habitat, in addition to other factors that are unknown or not understood. In desert tortoises (*Gopherus agassizii*), physiological and behavioral plasticity is key to survival during drought. These animals allow temporary increases in the osmotic and ionic concentrations of blood (Peterson 1996a), and they can acquire surpluses of energy while the water content and dry mass of the body are declining by eating grass that is low in water and protein during summer and autumn (Nagy and Medica 1986; Peterson 1996b). During drought, desert tortoises spend more time in burrows, which reduces the time spent out foraging, mating, or fighting. The retreat to burrows reduces both water and energy expenditure that enhances survival during periods of harsher desert conditions (Nagy and Medica 1986; Henen et al. 1998). The physiological and behavioral flexibility of desert tortoises appears to be centrally important to survival in harsh desert conditions when prolonged or severe drought can trigger die-offs in desert populations (Longshore et al. 2003).

Rainfall is the single climatic variable that explains much of the variation of energy acquisition and expenditure in desert tortoises, both directly in relation to free-standing water for drinking and indirectly through effects on the availability and quality of food (Peterson 1996b). Evidently, free water is necessary for achieving a net annual profit of energy. Similar freshwater dependence characterizes the Gila monster lizard (*Heloderma suspectum*) in the Sonoran Desert where rainfall influences the timing and duration of surface activity, the uses and storage of energy, and tolerance of physiological disturbances to endure seasonal limitations of resources (Davis and DeNardo 2010). These lizards also utilize the urinary bladder as a long-term water reservoir and exhibit "binge" drinking behavior that can increase body mass and storage up to nearly 22% while reducing plasma osmolality by 24% within 24 h of access to water (Davis and DeNardo 2007). Despite drought and the seasonal availability of resources, Gila monsters are able to capitalize on pulsatile energy resources as well as manage their storage to support growth and reproduction while enduring seasonal limitations of resources (Davis and DeNardo 2010). The importance of free water in desert reptiles can be seen from experimental supplementation of water to influence the behavioral ecology of Gila monsters. Lizards that had periodic supplementation of water during seasonal drought were active above ground for significantly greater proportions of time than were controls (Davis and DeNardo 2009). The increases in surface activity enhanced the acquisition of food and led to larger stores of energy compared with controls during 2 years of the study.

Reptiles exhibit specialized behaviors for collecting water in arid or seasonally dry habitats (Lasiewski and Bartholomew 1969; Malik et al. 2014; and references therein). Microsculpturing features of the stratum corneum of integument have been studied as interesting and important elements in the collection of airborne moisture in the form of rainfall, fog, mist, and dew in several species of reptiles (Bentley and Blumer 1962; Gans et al. 1982; Schwenk and Greene 1987; Lillywhite and Sanmartino 1993; Andrade and Abe 2000; Comanns et al. 2011) as well as amphibians (Lillywhite and Licht 1974; Tracy et al. 2011). It is highly probable that specializations of skin and behavior for harvesting free water will continue to be discovered in a variety of taxa living in arid habitats.

The ability of desert animals to thrive in the face of extreme heat, scarcity of free water, and uncertainty has attracted much interest and investigation by researchers working at various levels of inquiry. Here, I stress the viewpoint that novel mechanisms and phenomena underlying evolutionary adaptation to extreme environments will likely be a part of continued discoveries when creative approaches to research are aimed at better understanding the survival of amphibians and reptiles (Lillywhite and Navas 2006). A recent study of a desert rodent illustrates the point. The spinifex hopping mouse (*Notomys alexis*) can maintain water balance without drinking water, partly because of savings attributable to an efficient kidney that produces small volumes of highly concentrated urine. Little was known until recently regarding how obligatory losses of water were compensated by input. Takei et al. (2012) investigated this question and found that depriving the mouse of water induced a higher level of food intake, driven by changes in plasma leptin and ghrelin and the parallel expression of neuropeptides in the hypothalamus that stimulate appetite. As the deprivation of water was prolonged, body mass increased gradually because of hepatic glycogen storage following an initial period of catabolism of body fat. The metabolic strategy of the mouse switched during water deprivation from lipids to carbohydrates, which enhances metabolic water production per oxygen molecule. This mechanism also minimizes respiratory water loss. These changes in appetite regulation and energy metabolism were absent or less prominent in laboratory mice, and therefore could be important for survival of desert rodents in xeric environments. It seems that molecular studies related to water and energy metabolism might be profitable lines of inquiry in amphibians and reptiles, including evolutionary and phylogenetic aspects as well as physiological plasticity.

Importance of free water

Three sources of water are potentially available to animals: (1) dietary water, which is the free water available in food; (2) metabolic water that is formed during the metabolism of energy substrates derived from the digestive products of food; and (3) free water in the environment, for example, rainfall. Few animals are able to survive without the third source of water, and those that do generally are herbivores and eat plant material that is high in water content. However, the dependence or independence of animals with respect to this third water source is often either not known or misunderstood.

Various animals have been shown to rely heavily on dietary and metabolically produced water in xeric environments, and in some species, these totally satisfy water requirements (e.g., Nagy and Gruchacz 1994; Znari and Nagy 1997). Reliance on dietary water can influence foraging behaviors, and some species may shift to dietary items with greater water content when free water becomes increasingly limited (e.g., Nagy and Gruchacz 1994). Such options are most characteristic of herbivorous animals, however, and many carnivorous reptiles may not have such options. While the prey of many reptiles generally has water content from 60% to 75%, it is not clear whether consumption of prey can satisfy water requirements when free drinking water is scarce or absent. Quantitative assessment of this question is required considering that (1) foraging entails some amount of evaporative water loss associated with the activity; (2) digestion requires water; (3) excretion of metabolic wastes, especially from protein catabolism, requires water; (4) excretion of feces requires some water; and (5) secretion of salts using extrarenal salt glands also eliminates some water, albeit lesser amounts than might otherwise be excreted from the kidneys.

The central question of whether consumption of a meal incurs a net gain or loss of water was addressed in an important study using the desert Gila monster, *H. suspectum* (Wright et al. 2013). The investigation examined the short-term impact of meal consumption on osmolality of blood plasma and whether eating can maintain the hydration state over extended periods of time. Data indicate that a single meal incurs an acute short-term cost of water regardless of the status of hydration. Further, these lizards could not maintain water balance over long time scales if relying on meal consumption alone. The results of this study together with previous research (Davis and DeNardo 2007, 2009, 2010) demonstrate that Gila monsters are reliant on seasonal rainfall and require free-standing water to maintain water balance. Thus, some desert reptiles cannot acquire a net gain of water from their food. Further insights concerning the influence of meal consumption on water balance can be found in consideration of marine reptiles.

Marine snakes and reliance on free water

Marine snakes are instructive subjects for studies of osmoregulation, especially in the context of evolutionary transitions from terrestrial to marine environments (Dunson and Mazzotti 1989; Lillywhite 2014a). Their liquid environment has a high salt content, and their generally piscivorus diet entails consumption of prey with a relatively high content of protein. Moreover, marine snakes must remain active to exchange respiratory gases during bouts of air breathing at the ocean's surface, and they appear not to have the options of seclusion and dormancy that are available to terrestrial vertebrates. As in all marine reptiles, extrarenal salt glands (located beneath the tongue sheath in sea snakes) enable excretion of salts at high concentrations, and there seems little doubt that this capacity contributes to sea snakes being able to thrive in their salty environment. The rates of salt secretion from salt glands vary among species, and those with higher rates inhabit waters of greater salinity and tend to have more extensive oceanic ranges (Brischoux et al. 2012). However, salt glands do not appear to enable sea snakes to remain in water balance by drinking sea water.

Contrary to earlier views, a number of species of sea kraits (Laticaudinae), true sea snakes (Hydrophiini), and marine file snakes (Acrochordidae: *Acrochordus granulatus*) were recently shown to depend on environmental sources of fresh water to maintain water balance (Lillywhite et al. 2008, 2012, 2014a). These snakes will consume dilute brackish water, but do not drink water that is more than 30% seawater (sea kraits; Lillywhite et al. 2008), or in the case of a hydrophiine species, 50% seawater (*Hydrophis* (*Pelamis*) *platurus*; Lillywhite et al. 2012). Importantly, no sea snake that has been tested voluntarily drinks sea water.

So, how do sea snakes survive in their salty environment? Generally, sea snakes appear to be strongly resistant to, and tolerant of, dehydration. Recently, my colleagues and I demonstrated that the pelagic sea snake

H. (Pelamis) platurus dehydrates at sea and possibly survives 6–7 months of seasonal drought near the Guanacaste coast of Costa Rica (Lillywhite et al. 2014b). Presumably, the only source of free water available to these snakes is fresh or dilute brackish water that forms as a temporary "lens" at the ocean surface during large amounts of rainfall. Elsewhere, the distribution and abundance of sea kraits correlates with precipitation and the spatial distribution of freshwater springs and estuaries (Lillywhite et al. 2008; Lillywhite and Tu 2011).

It is possible that some marine snakes acquire a net gain of water from prey, but this hypothesis seems increasingly unlikely and requires further investigation. Theoretical considerations (Lillywhite et al. 2008) and recent investigations of desert reptiles (Wright et al. 2013) suggest that a net gain of water from prey is unlikely. Moreover, the water required for digestion and elimination of metabolic wastes associated with fish consumption suggests that even a net loss of water could result from feeding. It is important to note that marine snakes collected during dry seasons are thirsty, dehydrated, and will voluntarily drink fresh water even when observations indicate they had been feeding on fish during the period preceding measurements (Lillywhite et al. 2014a,b). Moreover, captive snakes increase drinking of free water following meal consumption and will cease feeding when dehydrated to a moderate (but not critical) stage of dehydration (unpublished observations; see also Lillywhite et al. 2014a; French 1956). Thus, it may be concluded that some marine snakes are dependent on fresh water, will dehydrate without it, and might not gain a water profit from the ingestion of prey.

Currently available data for all reptiles and amphibians suggest that the impact of food consumption on both energy and water balance varies among species, with implications for understanding the variation that is seen in foraging behavior. More data are required to further evaluate the costs and benefits related to foraging, acquisition of meals, and the interactions of feeding with water balance.

Perspective for the future

Energy and water will continue to be focal aspects of research relating to the physiology and behavior of ectothermic vertebrates. Justification for this statement arises first out of historical emphasis and interest in these subjects. Secondly, these popular areas of investigation are now nurtured by a widespread concern for important issues related to climatic change. In addition, advancements in molecular analytical tools are fueling a renaissance in phylogenetic systematics and evolutionary biology that impacts all other investigations of amphibians and reptiles. Indeed, molecular and statistical methods are both creating controversies and advancing changes in relationships and taxonomy at a rate that is sometimes too rapid to allow adequate evaluation and acceptance by the concerned community of scientists (e.g., Bortolus 2008; Guerra-García et al. 2008). Finally, new taxa continue to be discovered, and there is a plethora of species worldwide about which we know little, if any, natural history. These issues are important owing to direct effects on research in comparative studies and conservation management.

As new species are discovered, remote areas are accessed, and science advances in developing countries, species new to science will be increasingly studied. As we have found in the past, it is important that the fundamental biology and ecology of novel organisms be understood as an underpinning as well as contextual framework for understanding, and appropriately appreciating, new discoveries and "downstream" progress in modern evolutionary, molecular, and functional biology. Hence, a plea for appreciation of natural history continues to be important for most, if not all, readers of this book (e.g., Bartholomew 1986, Dayton 2003; Greene 2005; Tewksbury et al. 2014). Natural history stimulates and enriches the applied aspects of biology, and, by providing a fundamental knowledge of organisms, is indispensable to the conservation of biodiversity. Similarly, well-maintained collections of biological materials are, in my view, increasingly important, and they contribute innumerably to science and society (Suarez and Tsutsui 2004).

My emphasis on the importance of natural history, which here can be defined as descriptive ecology and ethology (Greene 2005), accompanies recognition that biological sciences are experiencing an ever-increasing availability of large data sets (so-called "big data"). These are growing rapidly in areas related to genomics, phylogenetics, physiology, ethology, evolutionary biology, climate, and other disciplines of potential relevance to physiological ecologists. Hence, there will be an increasing tendency to mine masses of archived data, to build or utilize probability models of underlying processes that might be of interest, and to add stochastic elements to deterministic models. However, there is no automated technique for distinguishing causation from correlation, and the number and complexity of potential false findings grow larger as data sets increase in size (Spiegelhalter 2014). Thus, the appropriate application of statistical methods will be essential for persons who are charting these territories, and I hope the community of comparative and integrative biologists will not lose focus on fundamental biological understanding for the sake of undue obsession with methodology and a potentially dogmatic adherence to favored "tools" that otherwise impede innovation in questions and new approaches. A clear and present danger is for some who are trained largely or exclusively in mathematics and statistics to be ignorant of the biology they explore, which can "miss" the understanding of outcomes and possibly lead to repetition of earlier work. Expertise and caution are especially advised, for example, in relation to current efforts to digitize character states that are described in legacy taxonomic descriptions (Cui 2012; Burleigh et al. 2013). I also caution about the pitfalls of "garbage in–garbage out" potentially applicable to "guesstimates" of "missing data" in large data sets, and the temptation to "tweak" the modeling process to produce a particular or "wanted" outcome. On the other hand, robust computational modeling can lead to new insights, questions, and understanding of phenomena. For example, modeling approaches based on systems biology are being used to enhance understanding cancer as an evolutionary process and thereby advance the treatment and (more importantly) prevention of cancer (Pepper et al. 2014). Such applications can be transformative and produce entirely new paradigms. Closer to our subject, behavioral and physiological models can be combined with spatial climatic and geographic data to predict some of the important impacts of global warming on ectothermic amphibians and reptiles (Kearney et al. 2009).

Although one could argue that much of the fundamental biology of ectotherms has been uncovered and is widely understood, reductionistic studies will continue to explore the important and interesting details of structure and function in amphibians and reptiles. New approaches will generate incentives for progress and will employ new methodologies related to imaging, molecular and biochemical advances, field instrumentation, and computerized analysis of data. I believe that such advances in research will continue to generate important bridges between physiology and behavior. One example of such active research is the "pregnant agenda" of neuroethology, including studies of taxonomic diversity in brain and behavior related to environment and evolution, and the ongoing assessment of the patterns of activity and assemblages of cells that enable recognition of important stimuli that evoke specific behaviors (Bullock 1999). Other research frontiers involving reductionism will include genetics and reproduction, the role of hormones and the effects of contaminants on behavior, emerging pathogens and the impacts of disease on physiology and behavior, and the interactions of energy and water requirements related to the real-time influence of climatic change (see also Chapters 6, 7, and 8). Technological innovations are also producing research bridges connecting physiology and behavior with paleontology (Eagle et al. 2011; O'Keefe and Chiappe 2011; Clabby 2014).

I will end with a plea for a thoughtful integration of advances in research, which will require (1) effective collaboration among teams of colleagues (Lélé and Norgaard 2005); (2) awareness of historical literature and earlier questions related to extant hypotheses; and (3) appreciation, adoption, and nurture of good natural history (Bartholomew 1986). In the framework of a collective enterprise, I hope that physiologists, ethologists, and ecologists do not become so lost within a tree that many opportunities vanish with a disappearing forest (McNab 2012). Strategic integration of physiology with behavior and ecology will utilize a suite of methodological tools and conceptual approaches, with increasing importance to addressing urgent environmental challenges (Cooke et al. 2013).

Acknowledgments

I am grateful to many persons who have contributed discussion, past collaborations, and stimulation of thinking about the concepts that are discussed in this chapter. I also thank Tobias Wang and an anonymous reviewer for comments that stimulated improvements in the manuscript. I am grateful to Denis Andrade for the invitation to contribute this chapter and for his patience and encouragement during the writing.

References

Adolph, S.C. and W.P. Porter. 1993. Temperature, activity, and lizard life histories. *Am. Nat.* 142:273–295.

Alpert, P. 2006. Constraints of tolerance: Why are desiccation-tolerant organisms so small or rare? *J. Exp. Biol.* 209:1575–1584.

Andrade, D.V. and A.S. Abe. 2000. Water collection by the body in a viperid snake, *Bothrops moojeni. Amphibia-Reptilia* 21:485–492.

Angilletta, M.J. Jr., A.F. Bennett, H. Guderley, C.A. Navas, F. Seebacher, and R.S. Wilson. 2006. Coadaptation: A unifying principle in evolutionary thermal biology. *Physiol. Biochem. Zool.* 79:282–294.

Angilletta, M.J. Jr., P.H. Niewiarowski, and C.A. Navas. 2002. The evolution of thermal physiology in ectotherms. *J. Therm. Biol.* 27:249–268.

Armitage, K.B. 2014. *Marmot Biology: Sociality, Individual Fitness, and Population Dynamics*. Cambridge: Cambridge University Press.

Babonis, L.S. and F. Brischoux. 2012. Perspectives on the convergent evolution of tetrapod salt glands. *Integr. Comp. Biol.* 52:245–256.

Barbeau, T.R. and H.B. Lillywhite. 2005. Body wiping behaviors associated with cutaneous lipids in hylid tree frogs of Florida. *J. Exp. Biol.* 208:2147–2156.

Bartholomew, G.A. 1982. Physiological control of body temperature. In Gans, C. and F.H. Pough (eds.), *Biology of the Reptilia*, Vol. 12, 167–211. New York: Academic Press.

Bartholomew, G.A. 1986. The role of natural history in contemporary biology. *BioScience* 36:324–329.

Beaupre, S.J. 1993. An ecological study of oxygen consumption in the mottled rock rattlesnake, *Crotalus lepidus lepidus,* and the black-tailed rattlesnake, *Crotalus molossus molossus,* from two populations. *Physiol. Zool.* 66:437–454.

Beaupre, S.J. 2002. Modeling time-energy allocation in vipers: Individual responses to environmental variation and implications for populations. In Schuett, G., M. Höggren, M.E. Douglas, and H.W. Greene (eds.), *Biology of the Vipers*, 463–481. Eagle Mountain, Utah: Eagle Mountain Publishing LC.

Beck, D.D. 1995. Ecology and energetics of three sympatric rattlesnake species in the Sonoran Desert. *J. Herpetol.* 29:211–223.

Beck, D.D. and C.H. Lowe. 1994. Resting metabolism of helodermatid lizards: Allometric and ecological relationships. *J. Comp. Physiol. B* 164:124–129.

Belinsky, A., R.A. Ackerman, R. Dmi'el, and A. Ar. 2004. Water in reptilian eggs and hatchlings. In Deeming, D.C. (ed.), *Reptilian Incubation: Environment, Evolution and Behaviour*, 125–141. Nottingham, UK: Nottingham University Press.

Bennett, A.F. 1980. The thermal dependence of lizard behaviour. *Anim. Behav.* 28:752–762.

Bennett, A.F. 1987. Interindividual variation: An underutilized resource. In Feder, M.E., A.F. Bennett, W.W. Burggren, and R.B. Huey (eds.), *New Directions in Physiological Ecology*, 147–169. Cambridge: Cambridge University Press.

Bentley, P.J. 1966. Adaptations of amphibians to arid environments. *Science* 152:619–623.

Bentley, P.J. and W. Blumer. 1962. Uptake of water by the lizard, *Moloch horridus*. *Nature* 194:699–700.

Bentley, P.J. and T. Yorio. 1979. Do frogs drink? *J. Exp. Biol.* 79:41–46.

Blaustein, A.R. and B.A. Bancroft. 2007. Amphibian population declines: Evolutionary considerations. *BioScience* 57:437–444.

Blaylock, L.A., R. Ruibal, and K. Plat-Aloia. 1976. Skin structure and wiping behavior of phyllomedusine frogs. *Copeia* 1976:283–295.

Bonnet, X. and F. Brischoux. 2008. Thirsty sea snakes forsake refuge during rainfall. *Austral Ecol.* 33:911–921.

Bonnet, X., A. Fizesan, and C.L. Michel. 2013. Shelter availability, stress level and digestive performance in the aspic viper. *J. Exp. Biol.* 216:815–822.

Bortolus, A. 2008. Error cascades in the biological sciences: The unwanted consequences of using bad taxonomy in ecology. *Ambio* 37:114–118.

Bovo, R.P., O.A.V. Marques, and D.V. Andrade. 2012. When basking is not an option: Thermoregulation of a viperid snake endemic to a small island in the South Atlantic of Brazil. *Copeia* 2012:408–418.

Bradshaw, S.D. 1986. *Ecophysiology of Desert Reptiles*. North Ryde, Australia: Academic Press Australia.

Brattstrom, B.H. 1963. A preliminary review of the thermal requirements of amphibians. *Ecology* 44:238–255.

Brekke, D.R., S.D. Hillyard, and R.W. Winokur. 1991. Behavior associated with the water absorption response by the toad, *Bufo punctatus*. *Copeia* 1991:393–401.

Brischoux, F., X. Bonnet, and R. Shine. 2010. Conflicts between feeding and reproduction in amphibious snakes (sea kraits, *Laticauda* spp.). *Austral. Ecol.* 36:46–52.

Brischoux, F., R. Tingley, R. Shine, and H.B. Lillywhite. 2012. Salinity influences the distribution of marine snakes: Implications for evolutionary transitions to marine life. *Ecography* 35:994–1003.

Brown, G.P. and R. Shine. 2005. Do changing moisture levels during incubation influence phenotypic traits of hatchling snakes (*Tropidonophis mairii*, Colubridae)? *Physiol. Biochem. Zool.* 78:524–530.

Bruton, M.J., C.A. McAlpine, A.G. Smith, and C.E. Franklin. 2014. The importance of underground shelter resources for reptiles in dryland landscapes: A woma python case study. *Austral Ecol.* 39:819–829.

Bullock, T.H. 1999. Neuroethology has pregnant agendas. *J. Comp. Physiol. A* 185:291–295.

Burleigh, J.G., K. Alphonse, A.J. Alverson et al. 2013. Next-generation phenomics for the tree of life. *PLoS Currents Tree of Life*, June 26. Edition 1. Doi:10.1371/currents.tol.085c713acafc8711b2ff7010a4b03733.

Bush, M.B. 2002. Distributional change and conservation on the Andean flank, a palaeoecological perspective. *Global Ecol. Biogeogr.* 11:463–473.

Carvalho, J.E., C.A. Navas, and I.C. Pereira. 2010. Energy and water in aestivating amphibians. In Navas, C.A. and J.E. Carvalho (eds.), *Aestivation: Molecular and Physiological Aspects*, Progress in Molecular and Subcellular Biology, Vol. 49, 141–169. Heidelberg: Springer-Verlag.

Chamaillé-Jammes, S., M. Massott, P. Aragon, and J. Clobert. 2006. Global warming and positive fitness response in mountain populations of common lizards, *Lacerta vivipara*. *Global Change Biol.* 12:392–402.

Chen, I.-C., J.K. Hill, R. Ohlemüller, D.B. Roy, and C.D. Thomas. 2011. Rapid range shifts of species associated with high levels of climate warming. *Science* 333:1024–1026.

Christian, K. and D. Parry. 1997. Reduced rates of water loss and chemical properties of skin secretions of the frogs *Litoria caerulea* and *Cyclorana australis*. *Aust. J. Zool.* 45:13–20.

Christian, K.A., L.K. Corbett, B. Green, and B.W. Weavers. 1995. Seasonal activity and energetics of two species of varanid lizards in tropical Australia. *Oecologia* 103:349–357.

Christian, K.A., J.K. Webb, and T.J. Schultz. 2003. Energetics of bluetongue lizards (*Tiliqua scincoides*) in a seasonal tropical environment. *Oecologia* 136:515–523.

Clabby, C. 2014. Paleontology's x-ray excavations. *Am. Sci.* 102:386–388.

Comanns, P., C. Effertz, F. Hischen, K. Staudt, W. Böhme, and W. Baumgartner. 2011. Moisture harvesting and water transport through specialized microstructures on the integument of lizards. *Beilstein J. Nanotechnol.* 2:204–214.

Cooke, S.J., D.T. Blumstein, R. Bucholz et al. 2014. Physiology, behavior and conservation. *Physiol. Biochem. Zool.* 87:1–14.

Cooke, S.J., L. Sack, C.E. Franklin, A.P. Farrell, J. Beardall, M. Wikelski, and S.L. Chown. 2013. What is conservation physiology? Perspectives on an increasingly integrated and essential science. *Conserv. Physiol.* 1. DOI:10.1093/conphys/cot001.

Cowles, R.B. 1939. Possible implications of reptilian thermal tolerance. *Science* 90:465–466.

Cowles, R.B. and C.M. Bogert. 1944. A preliminary study of the thermal requirements of desert reptiles. *Bull. Am. Mus. Nat. Hist.* 83:261–296.

Cui, H. 2012. CharaParser for fine-grained semantic annotation of organism morphological descriptions. *J. Assoc. Inf. Sci. Technol.* 63:738–754.

Dantzler, W.H. and E.J. Braun. 1980. Comparative nephron function in reptiles, birds, and mammals. *Am. J. Physiol.* 239:R197–R213.

Davis, J.R. and D.F. DeNardo. 2007. The urinary bladder as a physiological reservoir that moderates dehydration in a large desert lizard, the Gila Monster *Heloderma suspectum*. *J. Exp. Biol.* 210:1472–1480.

Davis, J.R. and D.F. DeNardo. 2009. Water supplementation affects the behavioral and physiological ecology of Gila monsters (*Heloderma suspectum*) in the Sonoran Desert. *Physiol. Biochem. Zool.* 82:739–748.

Davis, J.R. and D.F. DeNardo. 2010. Seasonal patterns of body condition, hydration state, and activity of Gila monsters (*Heloderma suspectum*) at a Sonoral Desert site. *J. Herpetol.* 44:83–93.

Dayton, P.K. 2003. The importance of the natural sciences to conservation. *Am. Nat.* 162:1–13.

Dimmitt, M.A. and R. Ruibal. 1980. Environmental correlates of emergence in spadefoot toads (*Scaphiopus*). *J. Herpetol.* 14:21–29.

Donnelly, M.A. 1989. Demographic effects of reproductive resource supplementation in a territorial frog, *Dendrobates pumilio. Ecol. Monogr.* 59:207–221.

Du, W. 2004. Water exchange of flexible-shelled eggs and its effect on hatchling traits in the Chinese skink, *Eumeces chinensis. J. Comp. Physiol. B Biochem. Syst. Environ. Physiol.* 174:489–493.

Dunson, W.A. and F.J. Mazzotti. 1989. Salinity as a limiting factor in the distribution of reptiles in Florida Bay: A theory for the estuarine origin of marine snakes and turtles. *Bull. Mar. Sci.* 44:229–244.

Eagle, R.A., T. Tütken, T.S. Martin et al. 2011. Dinosaur body temperatures determined from isotopic ($^{13}C–^{18}O$) ordering in fossil biominerals. *Science* 333:443–445.

Edgell, T.C., B.R. Lynch, G.C. Trussell, and A.R. Palmer. 2009. Experimental evidence for the rapid evolution of behavioral canalization in natural populations. *Am. Nat.* 174:434–440.

Elton, C.S. 1927. *Animal Ecology.* Chicago, Illinois: University of Chicago Press.

Etheridge, K. 1990. Water balance in estivating sirenid salamanders (*Siren lacertina*). *Herpetologica* 46:400–406.

Feder, M.E., A.F. Bennett, and R.B. Huey. 2000. Evolutionary physiology. *Annu. Rev. Ecol. Syst.* 31:315–341.

French, R.L. 1956. Eating, drinking, and activity patterns in *Peromyscus maniculatus sonoriensis. J. Mammal.* 37:74–79.

Gallagher, K.J., D.A. Morrison, R. Shine, and G.C. Grigg. 1983. Validation and use of ^{22}Na turnover to measure food intake in free-ranging lizards. *Oecologia* 60:76–82.

Gans, C., R. Merlin, and W.F.C. Blumer. 1982. The water-collecting mechanism of *Moloch horridus* re-examined. *Amphibia-Reptilia* 3:57–64.

Garland, T. Jr. 1988. Genetic basis of activity metabolism—I. Inheritance of speed, stamina, and antipredator displays in the garter snake *Thamnophis sirtalis. Evolution* 42:335–350.

Garland, T. Jr. and P.A. Carter. 1994. Evolutionary physiology. *Annu. Rev. Physiol.* 56:579–621.

Garland, T. Jr., R.B. Huey, and A.F. Bennett. 1991. Phylogeny and coadaptation of thermal physiology in lizards: A reanalysis. *Evolution* 45:1969–1975.

Garland, T. Jr. and M.R. Rose. 2009. *Experimental Evolution: Concepts, Methods and Applications of Selection Experiments.* Berkeley, California: University of California Press.

Gatten, R.E. Jr. 1987. Activity metabolism of anuran amphibians: Tolerance to dehydration. *Physiol. Zool.* 60:576–585.

Gienger, C.M., C.R. Tracy, and K.A. Nagy. 2014. Life in the lizard slow lane: Gila monsters have low rates of energy use and water flux. *Copeia* 2014:279–287.

Gifford, M.E., T.A. Clay, and R. Powell. 2012. Habitat use and activity influence thermoregulation in a tropical lizard, *Ameiva exsul. J. Thermal Biol.* 37:496–501.

Gomez, N.S., M. Acosta, F. Zaidan III, and H.B. Lillywhite. 2006. Wiping behavior, skin resistance, and the metabolic response to dehydration in the arboreal frog *Phyllomedusa hypochondrialis. Physiol. Biochem. Zool.* 79:1058–1068.

Greene, H.W. 2005. Organisms in nature as a central focus for biology. *Trends Ecol. Evol.* 20:23–27.

Gregory, P.T. 1982. Reptilian hibernation. In Gans, C. and F.H. Pough (eds.), *Biology of the Reptilia*, Vol. 13, 53–154. New York: Academic Press.

Guerra-García, J.M., F. Espinosa, and J.C. García-Gómez. 2008. Trends in taxonomy today: An overview about the main topics in taxonomy. *Zool. Baetica* 19:15–49.

Guppy, M. and P. Withers. 1999. Metabolic depression in animals: Physiological perspectives and biochemical generalizations. *Biol. Rev.* 74:1–40.

Harlow, P. and G. Grigg. 1984. Shivering thermogenesis in a brooding diamond python, *Python spilotes spilotes*. *Copeia* 1984:959–965.

Hazard, L.C., C. Lechuga, and S. Zilinskis. 2010. Secretion by the nasal salt glands of two insectivorous lizard species is initiated by an ecologically relevant dietary ion, chloride. *J. Exp. Zool.* 313A:442–451.

Henen, B.T., C.C. Peterson, I.R. Wallis, K.H. Berry, and K.A. Nagy. 1998. Effects of climatic variation on field metabolism and water relations of desert tortoises. *Oecologia* 117:365–373.

Hertz, P.E. 1992. Temperature regulation in Puerto Rican *Anolis* lizards: A field test using null hypotheses. *Ecology* 73:1405–1417.

Hickling, R., D.B. Roy, J.K. Hill, R. Fox, and C.D. Thomas. 2006. The distributions of a wide range of taxonomic groups are expanding polewards. *Global Change Biol.* 12:450–455.

Hildebrandt, J.P. 2001. Coping with excess salt: Adaptive functions of extrarenal osmoregulatory organs in vertebrates. *Zoology* 104:209–220.

Hillman, S.S. 1982. The effects of *in vivo* and *in vitro* hyperosmolality on skeletal muscle performance in the amphibians *Rana pipiens* and *Scaphiopus couchii*. *Comp. Biochem. Physiol. A* 73:709–712.

Hillman, S.S. 1984. Inotropic influence of dehydration and hyperosmolal solutions on amphibian cardiac muscle. *J. Comp. Physiol.* 154:325–328.

Hillyard, S.D. 1976. The movement of soil water across the isolated amphibian skin. *Copeia* 1976:314–320.

Hochachka, P.W. and G.N. Somero. 2002. *Biochemical Adaptation: Mechanism and Process in Physiological Evolution*. Oxford and New York: Oxford University Press.

Hoff, K von S. and S.D. Hillyard. 1991. Angiotensin II stimulates cutaneous drinking in the toad *Bufo punctatus*. *Physiol. Zool.* 64:1165–1172.

Hoffmann, A.A. and C.M. Sgrò. 2011. Climate change and evolutionary adaptation. *Nature* 470:479–485.

Huang, S.-P., C.-R. Chiou, T.-E. Lin, M.-C. Tu, C.-C. Lin, and W.P. Porter. 2013. Future advantages in energetics, activity time, and habitats predicted in a high-altitude pit viper with climate warming. *Funct. Ecol.* 27:446–458.

Huang, S.-P., W.P. Porter, M.-C. Tu, and C.-R. Chiou. 2014. Forest cover reduces thermally suitable habitats and affects responses to a warmer climate predicted in a high-elevation lizard. *Oecologia* 175:25–35.

Huey, R.B. 1982. Temperature, physiology, and the ecology of reptiles. In Gans, C. and F.H. Pough (eds.), *Biology of the Reptilia*, Vol. 12, 25–74. New York: Academic Press.

Huey, R.B., C.A. Deutsch, J.J. Tewksbury et al. 2009. Why tropical forest lizards are vulnerable to climate warming. *Proc. R. Soc. B* 276:1939–1948.

Huey, R.B., M.R. Kearney, A. Krockenberger, J.A. Holtum, M. Jess, and S.E. Williams. 2012. Predicting organismal vulnerability to climate warming: Roles of behavior, physiology and adaptation. *Philos. Trans. R. Soc. B* 367:1665–1679.

Huey, R.B. and M. Slatkin. 1976. Cost and benefits of lizard thermoregulation. *Quart. Rev. Biol.* 51:363–384.

Huey, R.B. and R.D. Stevenson. 1979. Integrating thermal physiology and ecology of ectotherms: A discussion of approaches. *Am. Zool.* 19:357–366.

Hutchison, V.H. and R.K. Dupre. 1992. Thermoregulation. In Feder, M.E. and W.W. Burggren (eds.), *Environmental Physiology of the Amphibians*, 206–249. Chicago, IL: University of Chicago Press.

Inger, R.F. 1959. Temperature responses and ecological relations of two Bornean lizards. *Ecology* 40:127–136.

Jones, F.C., M.G. Grabherr, Y.F. Chan et al. 2012. The genomic basis of adaptive evolution in threespine sticklebacks. *Nature* 484:55–61.

Kayes, S.M., R.L. Cramp, and C.E. Franklin. 2009. Metabolic depression during aestivation in *Cyclorana alboguttata*. *Comp. Biochem. Physiol. A* 154:557–563.

Kearney, M. and W. Porter. 2009. Mechanistic niche modeling: Combining physiological and spatial data to predict species ranges. *Ecol. Lett.* 12:334–350.

Kearney, M., R. Shine, and W.P. Porter. 2009. The potential for behavioral thermoregulation to buffer "cold-blooded" animals against climate warming. *Proc. Natl. Acad. Sci. USA* 106:3835–3840.

Kingsolver, J.G. and R.B. Huey. 1998. Evolutionary analyses of morphological and physiological plasticity in thermally variable environments. *Am. Zool.* 38:545–560.

Klauber, L.M. 1939. Studies of reptile life in the arid Southwest. *Bull. Zool. Soc. San Diego* 14:1–100.

Lande, R. 2009. Adaptation to an extraordinary environment by evolution of phenotypic plasticity and genetic assimilation. *J. Evol. Biol.* 22:1435–1446.

Lasiewski, R.C. and G.A. Bartholomew. 1969. Condensation as a mechanism for water gain in nocturnal desert poikilotherms. *Copeia* 1969:405–407.

Lélé, S. and R.B. Norgaard. 2005. Practicing interdisciplinarity. *BioScience* 55:967–975.

Lillywhite, H.B. 1987. Temperature, energetics, and physiological ecology. In Seigel, R.A., J.T. Collins, and S.S. Novak (eds.), *Snakes: Ecology and Evolutionary Biology*, 422–477. New York: Macmillan Publishing Company.

Lillywhite, H.B. 2006. Water relations of tetrapod integument. *J. Exp. Biol.* 209:202–226.

Lillywhite, H.B. 2013. Climate change and reptiles. In Rohde, K. (ed.), *The Balance of Nature and Human Impact*, 279–294. Cambridge: Cambridge University Press.

Lillywhite, H.B. 2014a. A desert in the sea: Water relations of marine snakes. In Lutterschmidt, W. (ed.), *Reptiles in Research: Investigations of Ecology, Physiology and Behavior from Desert to Sea*, 485–500. Hauppauge, NY: Nova Science Publishers.

Lillywhite, H.B. 2014b. *How Snakes Work: Structure, Function and Behavior of the World's Snakes*. New York: Oxford University Press.

Lillywhite, H.B., L.S. Babonis, C.M. Sheehy III, and M.-C. Tu. 2008. Sea snakes (*Laticauda* spp.) require fresh drinking water: Implication for the distribution and persistence of populations. *Physiol. Biochem. Zool.* 81:785–796.

Lillywhite, H.B., F. Brischoux, C.M. Sheehy III, and J.B. Pfaller. 2012. Dehydration and drinking responses in a pelagic sea snake. *Integr. Comp. Biol.* 52:227–234.

Lillywhite, H.B., H. Heatwole, and C.M. Sheehy III. 2014a. Dehydration and drinking behavior of the marine file snake, *Acrochordus granulatus*. *Physiol. Biochem. Zool.* 87:46–55.

Lillywhite, H.B. and P. Licht. 1974. Movement of water over toad skin: Functional role of epidermal sculpturing. *Copeia* 1974:165–171.

Lillywhite, H.B., P. Licht, and P. Chelgren. 1973. The role of behavioral thermoregulation in the growth energetics of the toad, *Bufo boreas. Ecology* 54:375–383.

Lillywhite, H.B. and A.K. Mittal. 1999. Amphibian skin and the aquatic-terrestrial transition: Constraints and compromises related to water exchange. In Mittal, A.K., F.B. Eddy, and J.S. Datta Munshi (eds.), *Water/Air Transition in Biology*, 131–144. Enfield, NH: Science Publishers.

Lillywhite, H.B., A.K. Mittal, T.K. Garg, and N. Agrawal. 1997a. Integumentary structure and its relationship to wiping behavior in the common Indian tree frog, *Polypedates maculatus. J. Zool. (Lond.)* 243:675–687.

Lillywhite, H.B., A.K. Mittal, T.K. Garg, and N. Agrawal. 1997b. Wiping behavior and its ecophysiological significance in the Indian tree frog, *Polypedates maculatus. Copeia* 1997:88–100.

Lillywhite, H.B. and C.A. Navas. 2006. Animals, energy, and water in extreme environments: Perspectives from Ithala 2004. *Physiol. Biochem. Zool.* 79:265–273.

Lillywhite, H.B. and V. Sanmartino. 1993. Permeability and water relations of hygroscopic skin of the file snake, *Acrochordus granulatus. Copeia* 1993:99–103.

Lillywhite, H.B., C.M. Sheehy III, F. Brischoux, and A. Grech. 2014b. Pelagic sea snakes dehydrate at sea. *Proc. R. Soc. B* 281: 20140119.

Lillywhite, H.B. and M.-C. Tu. 2011. Abundance of sea kraits correlates with precipitation. *PLoS ONE* 6(12):e28556. DOI:10.1371/journal.pone.0028556.

Longshore, K.M., J.R. Jaeger, and J.M. Sappington. 2003. Desert tortoise (*Gopherus agassizii*) survival at two Eastern Mojave Desert sites: Death by short-term drought? *J. Herpetol.* 37:169–177.

Luiselli, L. and G.C. Akani. 2002. Is thermoregulation really unimportant for tropical reptiles? Comparative study of four sympatric snake species from Africa. *Acta Oecol.* 23:59–68.

Lukoschek, V. and J.S. Keogh. 2006. Molecular phylogeny of sea snakes reveals a rapidly diverged adaptive radiation. *Biol. J. Linn. Soc.* 89:523–539.

Maderson, P.F.A. 1972. When? Why? And how? Some speculations on the evolution of vertebrate integument. *Am. Zool.* 12:159–171.

Madsen, T. and R. Shine. 1993. Costs of reproduction in a population of European adders. *Oecologia* 94:488–495.

Malik, F.T., R.M. Clement, D.T. Gethin, W. Krawszik, and A.R. Parker. 2014. Nature's moisture harvesters: A comparative review. *Bioinspir. Biomim.* 9:031002. DOI:10.1088/1748-3182/9/3/031002.

McArthur, L.J. 2007. The seasonal energetics of three species of Australian tropical frogs (Anura: Hylidae). Doctoral thesis, Darwin, Australia: Charles Darwin University.

McClanahan, L. Jr. and R.A. Baldwin. 1969. Rate of water uptake through the integument of the desert toad, *Bufo punctatus. Comp. Biochem. Physiol.* 28:381–389.

McClanahan, L. Jr., J.N. Stinner, and V.H. Shoemaker. 1978. Skin lipids, water loss, and energy metabolism in a South American tree frog (*Phyllomedusa sauvagei*). *Physiol. Zool.* 51:179–187.

McCue, M.D. 2010. Starvation physiology: Reviewing the different strategies animals use to survive a common challenge. *Comp. Biochem. Physiol.* 156A:1–18.

McCue, M.D., H.B. Lillywhite, and S.J. Beaupre. 2012. Physiological responses to starvation in snakes: Low energy specialists. In McCue, M.D. (ed.), *Comparative Physiology of Fasting, Starvation, and Food Limitation*, 103–131. Heidelberg: Springer.

McNab, B.K. 2002. *The Physiological Ecology of Vertebrates: A View from Energetics.* Ithaca, NY: Cornell University Press.

McNab, B.K. 2012. *Extreme Measures: The Ecological Energetics of Birds and Mammals.* Chicago, Illinois: The University of Chicago Press.

Meiri, S., A.M. Bauer, L. Chirio et al. 2013. Are lizards feeling the heat? A tale of ecology and evolution under two temperatures. *Global Ecol. Biogeogr.* 22:834–845.

Milsom, W.K., C. Sanders, C. Leite, A.S. Abe, D.V. Andrade, and G. Tattersall. 2012. Seasonal changes in thermoregulatory strategies of Tegu lizards. In Ruf, T., C. Bieber, W. Arnold, and E. Millesi (eds.), *Living in a Seasonal World*, 317–324. Heidelberg: Springer-Verlag.

Mitcheletti, S., E. Parra, and E.J. Routman. 2012. Adaptive color polymorphism and unusually high local genetic diversity in the side-blotched lizard, *Uta stansburiana. PLoS ONE* 7(10):e47694. DOI:10.1371/journal.pone.0047694.

Mitchell, N.J., M.R. Kearney, N.J. Nelson, and W.P. Porter. 2008. Predicting the fate of a living fossil: How will global warming affect sex determination and hatching phenology in tuatara? *Proc. R. Soc. Lond. B* 275:2185–2193.

Moore, I.T. and T.S. Jessop. 2003. Stress, reproduction, and adrenocortical modulation in amphibians and reptiles. *Horm. Behav.* 43:39–47.

Mosauer, W. 1936. The toleration of solar heat in desert reptiles. *Ecology* 17:56–66.

Mosauer, W. and E.L. Lazier. 1933. Death from insolation in desert snakes. *Copeia* 1933:149.

Nagy, K.A. 1983. Ecological energetics. In Huey, R.B., E.R. Pianka, and T.W. Schoener (eds.), *Lizard Ecology*, 24–54. Cambridge: Harvard University Press.

Nagy, K.A. and M.J. Gruchacz. 1994. Water and energy metabolism of the desert-dwelling kangaroo rat (*Dipodomys merriami*). *Physiol. Zool.* 67:1461–1478.

Nagy, K.A. and P.A. Medica. 1986. Physiological ecology of desert tortoises in southern Nevada. *Herpetologica* 42:73–92.

Nagy, K.A., M.K. Seely, and R. Buffenstein. 1993. Surprisingly low field metabolic rate of a diurnal desert gecko, *Rhoptropus afer. Copeia* 1993:216–219.

Nagy, K.A. and V.H. Shoemaker. 1975. Energy and nitrogen budgets of the free-living lizard *Sauromalus obesus. Physiol. Zool.* 48:252–262.

Navas, C.A., M.M. Antoniazzi, J.E. Carvalho, H. Suzuki, and C. Jared. 2007. Physiological basis for diurnal activity in dispersing juvenile *Bufo granulosus* in the Caatinga, a Brazilian semi-arid environment. *Comp. Biochem. Physiol.* 147A:647–657.

O'Connor, M.P., A.E. Sieg, and A.E. Dunham. 2006. Linking physiological effects on activity and resource use to population level phenomena. *Integr. Comp. Biol.* 46:1093–1109.

O'Keefe, F.R. and L.M. Chiappe. 2011. Viviparity and K-selected life history in a Mesozoic marine plesiosaur (Reptilia, Sauropterygia). *Science* 333:870–873.

O'Leary, T. and E. Marder. 2014. Mapping neural activation onto behavior in an entire animal. *Science* 344:372–373.

Peaker, M. and J.L. Linzell. 1975. *Salt Glands in Birds and Reptiles.* Cambridge: Cambridge University Press.

Pearse, A.S. 1931. *Animal Ecology*. New York: McGraw-Hill.

Pepper, J.W., B.K. Dunn, R.M. Fagerstrom, J.K. Gohagan, and N.A. Vydelingum. 2014. Using systems biology to understand cancer as an evolutionary process. *J. Evol. Med.* 2:1–8.

Peterson, C.C. 1996a. Anhomeostasis: Seasonal water and solute relations in two populations of the desert tortoise (*Gopherus agassizii*) during chronic drought. *Physiol. Zool.* 69:1324–1358.

Peterson, C.C. 1996b. Ecological energetics of the desert tortoise (*Gopherus agassizii*): Effects of rainfall and drought. *Ecology* 77:1831–1844.

Porter, W.P. and D.M. Gates. 1969. Thermodynamic equilibria of animals with environment. *Ecol. Monogr.* 39:245–270.

Porter, W.P. J.W. Mitchell, W.A. Beckman, and C.B. DeWitt. 1973. Behavioral implications of mechanistic ecology: Thermal and behavioral modeling of desert ectotherms and their microenvironment. *Oecologia* 13:1–54.

Pough, F. H. 1974. Preface to the republication of *A Preliminary Study of the Thermal Requirements of Desert Reptiles* by R.B. Cowles and C.M. Bogert, i–iv. Athens, Ohio: Society for the Study of Amphibians and Reptiles.

Pough, F.H. 1980. The advantages of ectothermy for tetrapods. *Am. Nat.* 115:92–112.

Pough, F.H. 1983. Amphibians and reptiles as low-energy systems. In Aspey, W.P. and S.I. Lustick (eds.), *Behavioral Energetics: The Cost of Survival in Vertebrates*, 141–188. Columbus, OH: Ohio State University Press.

Pounds, J.A., M.R. Bustamante, L.A. Coloma et al. 2006. Widespread amphibian extinctions driven by global warming. *Nature* 439:161–167.

Prates, I. and C.A. Navas. 2009. Cutaneous resistance to evaporative water loss in Brazilian *Rhinella* (Anura: Bufonidae) from contrasting environments. *Copeia* 2009:618–622.

Preest, M.R. and F.H. Pough. 2003. Effects of body temperature and hydration state on organismal performance of toads, *Bufo americanus*. *Physiol. Biochem. Zool.* 76:229–239.

Qian, H. 2010. Environment-richness relationships for mammals, birds, reptiles, and amphibians at global and regional scales. *Ecol. Res.* 25:629–637.

Raxworthy, C.J., R.G. Pearson, N. Rabibisoa et al. 2008. Extinction vulnerability of tropical montane endemism from warming and upslope displacement: A preliminary appraisal for the highest massif in Madagascar. *Global Change Biol.* 14:1703–1720.

Reading, C.J., L.M. Luiselli, G.C. Akani et al. 2010. Are snake populations in widespread decline? *Biol. Lett.* 6:777–780.

Reilly, B.D., D.I. Schlipalius, R.L. Crump, P.R. Ebert, and C.E. Franklin. 2013. Frogs and estivation: Transcriptional insights into metabolism and cell survival in a natural model of extended muscle disuse. *Physiol. Genomics* 45:377–388.

Reno, H.W., F.R. Gehlbach, and R.A. Turner. 1972. Skin and aestivational cocoon of the aquatic amphibian, *Siren intermedia* Le Conte. *Copeia* 1972:625–631.

Reynolds, S.J., K.A. Christian, and C.R. Tracy. 2010. The cocoon of the fossorial frog *Cyclorana australis* functions primarily as a barrier to water exchange with the substrate. *Physiol. Biochem. Zool.* 83:877–884.

Rohde, K. (ed.) 2013. *The Balance of Nature and Human Impact*. Cambridge: Cambridge University Press.

Rohner, N., D.F. Jarosz, J.E. Kowalko et al. 2013. Cryptic variation in morphological evolution: HSP90 as a capacitor for loss of eyes in cavefish. *Science* 342:1372–1375.

Rosenblum, E.B. 2006. Convergent evolution and divergent selection: Lizards at the White Sands ecotone. *Am. Nat.* 167:1–15.

Ruibal, R. 1961. Thermal relations of five species of tropical lizards. *Evolution* 15:98–111.

Ruibal, R. and S.S. Hillman. 1981. Cocoon structure and function in the burrowing hylid frog, *Pternohyla fodiens*. *J. Herpetol.* 15:403–408.

Ruibal, R., L. Tevis, and V. Roig. 1969. The terrestrial ecology of the spadefoot toad *Scaphiopus hammondi*. *Copeia* 1969:571–584.

Rull, V. and T. Vegas-Vilarrúbia. 2006. Unexpected biodiversity loss under global warming in the neotropical Guayana Highlands: A preliminary appraisal. *Global Change Biol.* 12:1–9.

Sanders, K.L., A.R. Rasmussen, Mumpuni et al. 2013. Recent rapid speciation and ecomorph divergence in Into-Australian sea snakes. *Mol. Ecol.* 22:2742–2759.

Schoener, T.W. 1974. Resource partitioning in ecological communities. *Science* 185:27–39.

Schwenk, K. and H.W. Greene. 1987. Water collection and drinking in *Phrynocephalus helioscopus*: A possible condensation mechanism. *J. Herpetol.* 21:134–139.

Secor, S.M. 1995. Ecological aspects of foraging mode for the snakes *Crotalus cerastes* and *Masticophis flagellum*. *Herpetol. Monogr.* 9:169–186.

Secor, S.M. and K.A. Nagy. 1994. Energetic correlates of foraging mode for the snakes *Crotalus cerastes* and *Masticophis flagellum*. *Ecology* 75:1600–1614.

Seebacher, F. and R.A. Alford. 1999. Movements and microhabitat use of a terrestrial amphibian (*Bufo marinus*) on a tropical island: Seasonal variation and environmental correlates. *J. Herpetol.* 33:208–214.

Seebacher, F., C.R. White, and C.E. Franklin. 2015. Physiological plasticity increases resilience of ectothermic animals to climate change. *Nat. Clim. Change* 5:61–66.

Seymour, R.S. 1973. Energy metabolism of dormant spadefoot toads (*Scaphiopus*). *Copeia* 1973:435–445.

Shine, R. and M.B. Thompson. 2006. Did embryonic responses to incubation conditions drive the evolution of reproductive modes in squamate reptiles? *Herpetol. Monogr.* 20:159–171.

Shoemaker, V.H., D. Balding, R. Ruibal, and L. McClanahan Jr. 1972. Uricotelism and low evaporative water loss in a South American frog. *Science* 175:1018–1020.

Shoemaker, V.H. and P.E. Bickler. 1979. Kidney and bladder function in a uricotelic treefrog (*Phyllomedusa sauvagei*). *J. Comp. Physiol. B* 133:211–218.

Shoemaker, V.H., L. McClanahan, P.C. Withers, S.S. Hillman, and R.C. Drewes. 1987. Thermoregulatory response to heat in the waterproof frogs *Phyllomedusa* and *Chiromantis*. *Physiol. Zool.* 60:365–372.

Sinervo, B., F. Méndez-de-la-Cruz, D.B. Miles et al. 2010. Erosion of lizard diversity by climate. *Science* 328:894–899.

Snyder, G.K. 1971. Adaptive value of a reduced respiratory metabolism in a lizard. A unique case. *Respir. Physiol.* 18:90–101.

Somero, G.N., E. Dahlhoff, and J.J. Lin. 1996. Stenotherms and eurytherms: Mechanisms establishing thermal optima and tolerance ranges. In Johnston, I.A. and A.F. Bennett (eds.), *Animals and Temperature*, 53–78. Cambridge: Cambridge University Press.

Spiegelhalter, D.J. 2014. Statistics. The future lies in uncertainty. *Science* 345:264–265.

Stahlschmidt, Z. and D.F. DeNardo. 2010. Parental behavior in pythons is responsive to both the hydric and thermal dynamics of the nest. *J. Exp. Biol.* 213:1691–1696.

Stewart, M.M. and F.H. Pough. 1983. Population density of tropical forest frogs: Relation to retreat sites. *Science* 221:570–572.

Stille, W.T. 1951. The nocturnal amphibian fauna of the southern Lake Michigan beach. *Ecology* 33:149–162.

Stille, W.T. 1958. The water absorption response of an anuran. *Copeia* 1958:217–218.

Suarez, A.V. and N.D. Tsutsui. 2004. The value of museum collections for research and society. *BioScience* 54:66–74.

Takei, Y., R.C. Bartolo, H. Fujihara, Y. Ueta, and J.A. Donald. 2012. Water deprivation induces appetite and alters metabolic strategy in *Notomys alexis*: Unique mechanisms for water production in the desert. *Proc. R. Soc. B* 279:2599–2608.

Tewksbury, J.J., J.G.T. Anderson, J.D. Bakker et al. 2014. Natural history's place in science and society. *BioScience* 64:300–310.

Tewksbury, J.J., R.B. Huey, and C.A. Deutsch. 2008. Putting the heat on tropical animals. *Science* 320:1296–1297.

Titon, B. Jr., C.A. Navas, J. Jim, and F.R. Gomes. 2010. Water balance and locomotor performance in three species of neotropical toads that differ in geographical distribution. *Comp. Biochem. Physiol. A* 156:129–135.

Toledo, R.C. and C. Jared. 1995. Cutaneous granular glands and amphibian venoms. *Comp. Biochem. Physiol.* 111A:1–29.

Tracy, C.R. 1976. A model of the dynamic exchanges of water and energy between a terrestrial amphibian and its environment. *Ecol. Monogr.* 46:293–326.

Tracy, C.R., K.A. Christian, and C.R. Tracy. 2010. Not just small, wet, and cold: Effects of body size and skin resistance on thermoregulation and arboreality of frogs. *Ecology* 91:1477–1484.

Tracy, C.R., N. Laurence, and K.A. Christian. 2011. Condensation onto the skin as a means of water gain by tree frogs in tropical Australia. *Am. Nat.* 178:553–558.

Tran, D.-Y., K. von, S. Hoff, and S.D. Hillyard. 1992. Effects of angiotensin II and bladder condition on hydration behavior and water uptake in the toad, *Bufo woodhousei*. *Comp. Biochem. Physiol.* 103A:127–130.

Turner, J.S. 1984. Raymond B. Cowles and the biology of temperature in reptiles. *J. Herpetol.* 18:421–436.

Turner, J.S. 1987. The cardiovascular control of heat exchange: Consequences of body size. *Am. Zool.* 27:69–79.

Ultsch, G.R. 1989. Ecology and physiology of hibernation and overwintering among freshwater fishes, turtles, and snakes. *Biol. Rev.* 64:453–516.

van Beurden, E.K. 1980. Energy metabolism of dormant Australian water-holding frogs (*Cyclorana platycephalus*). *Copeia* 1980:787–799.

Van Mierop, L.H.S. and S.M. Barnard. 1976. Thermoregulation in a brooding female *Python molurus bivittatus*. *Copeia* 1976:398–401.

Van Mierop, L.H.S. and S.M. Barnard. 1978. Further observations on thermoregulation in the brooding female *Python molurus bivittatus* (Serpentes: Boidae). *Copeia* 1978:615–621.

Vitt, L.J., T.C.S. Avila-Pires, J.P. Caldwell, and V.R.L. Oliveira. 1998. The impact of individual tree harvesting on thermal environments of lizards in Amazonian rain forest. *Conserv. Biol.* 12:654–664.

Vleck, D. 1991. Water economy and solute regulation in reptilian and avian embryos. In Deeming, D.C. and M.W.J. Ferguson (eds.), *Egg Incubation: Its Effect on Embryonic Development in Birds and Reptiles*, 245–260. Cambridge, Cambridge University Press.

Weese, A.O. 1919. Environmental reactions of *Phrynosoma*. *Am. Nat.* 53:33–54.

Withers, P.C. 1995. Cocoon formation and structure in the aestivating Australian desert frogs, *Neobatrachus* and *Cyclorana*. *Aust. J. Zool.* 43:429–441.

Withers, P.C. 1998. Evaporative water loss and the role of cocoon formation in Australian frogs. *Aust. J. Zool.* 46:405–418.

Withers, P.C., S.S. Hillman, and R.C. Drewes. 1984. Evaporative water loss and skin lipids of anuran amphibians. *J. Exp. Biol.* 232:11–17.

Withers, P.C., S.S. Hillman, R.C. Drewes, and O.M. Sokol. 1982. Water loss and nitrogen excretion in sharp-nosed reed frogs (*Hyperolius nasutus*: Anura, Hyperoliidae). *J. Exp. Zool.* 97:335–343.

Wright, C.D., M.L. Jackson, and D.F. DeNardo. 2013. Meal consumption is ineffective at maintaining or correcting water balance in a desert lizard, *Heloderma suspectum*. *J. Exp. Biol.* 216:1439–1447.

Wygoda, M.L. 1984. Low evaporative water loss in an arboreal frog. *Herpetologica* 44:251–257.

Young, J.E., K.A. Christian, S. Donnellan, C.R. Tracy, and D. Parry. 2005. Comparative analysis of cutaneous evaporative water loss in frogs demonstrates correlation with ecological habits. *Physiol. Biochem. Zool.* 78:847–856.

Young, J.E., C.R. Tracy, K.A. Christian, and L.J. McArthur. 2006. Rates of cutaneous evaporative water loss of native Fijian frogs. *Copeia* 2006:83–88.

Znari, M. and K.A. Nagy. 1997. Field metabolic rate and water flux in free-living Bibron's agama (*Agama impalearis*, Boettger, 1874) in Moroccco. *Herpetologica* 53:81–88.

chapter two

Acclimation, acclimatization, and seasonal variation in amphibians and reptiles

Alexander G. Little and Frank Seebacher

Contents

Introduction

Temperature can limit animal performance by constraining the biochemical reaction rates that support physiological functions at all levels. This limitation exists because temperature influences the free energy available to drive reactions forward. Temperature extremes can also compromise reaction rates by denaturing enzymes and altering membrane fluidity (Hazel 1995; O'Brien 2011). All animals therefore have thermal tolerance windows that are set by upper and lower critical temperature thresholds (CT_{min} and CT_{max}), outside of which the animal does not survive (Figure 2.1). Within these CT values lies the performance breadth, which comprises the range of body temperatures where performance exceeds, for example, 80% of maximal performance (Huey and Stevenson 1979; Huey and Kingsolver 1989; Angilletta et al. 2010; Huey et al. 2012). Performance breadth tends to be a more ecologically relevant measure

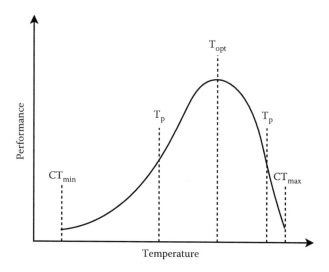

Figure 2.1 Thermal reaction norm. Animals operate within their pejus temperatures (Tp) and have optimal performance at T_{opt}. Performance declines as they approach critical temperatures (CT_{min} and CT_{max}), at which point death occurs.

than CT range because it represents the temperatures over which animals, populations, and species can maintain evolutionary fitness. As a result, it is this range, rather than CT_{min}–CT_{max}, that often determines population- and species-level geographic range limits (Pörtner 2001, 2002; Kearney and Porter 2009; Seebacher and Franklin 2011; McCann et al. 2014). For instance, recent work shows that cane toads (*Bufo marinus*) can lower their CT_{min} when exposed to cold temperatures for periods as short as 12 h (McCann et al. 2014). But the ecological relevance of this finding is questionable because locomotion becomes more metabolically costly at low temperatures and their capacity for extended movement is diminished (Seebacher and Franklin 2011; McCann et al. 2014; Seebacher et al. 2014). Thus, it is their performance breadth, not their CT range that constrains dispersal and range expansion into the relatively cooler regions across their southern range in Australia (Seebacher and Franklin 2011).

Amphibians and reptiles are ectotherms and therefore do not produce and retain sufficient metabolic heat to alter their body temperatures. Instead, body temperatures are determined by heat transfer with the environment (by radiation, convection, evaporation, and conduction). Without physiological or behavioral adjustments, the biochemical rate functions of ectotherms are more susceptible to changes in environmental temperatures. The Q_{10} temperature coefficient for biological systems is typically between 2 and 3, meaning that reaction rates will be more than halved for every 10°C drop in body temperature, and more than doubled for every 10°C rise in body temperature. As animals approach their thermal extremes, the growing mismatch between energy supply and demand ultimately collapses metabolic scope (the difference between basal metabolic rate and maximal metabolic rate; Figure 2.2; Pörtner 2001; Pörtner and Knust 2007). At low temperatures, rates of energy (adenosine triphosphate [ATP]) production fall more quickly than resting energy expenditure, until supply can no longer sustain demand (Pörtner 2001). At high temperatures, rates of energy expenditure rise exponentially and energy production begins to decrease, so that demand begins to exceed supply (Pörtner 2001; Pörtner and Knust 2007). Despite their dependence on environmental temperature, reptiles and amphibians have evolved to thrive in some of the most extreme thermal environments known, from wood frogs in the snowy tundra of the Arctic Circle to snakes and lizards in the dry heat of the Sahara Desert. In addition, many ectotherms can function normally for large parts of the year even though thermal conditions and body temperatures within individuals can vary considerably.

There are different approaches reptiles and amphibians use to cope with seasonal variation in temperature. Some conform to ambient temperatures and enter a depressed metabolic state (hypometabolism) to conserve energy stores until optimal or near-optimal conditions return (Boutilier et al. 1997; Storey and Storey 2004; Navas and Carvalho 2010). Other reptiles and amphibians compensate for seasonal changes in temperature so that physiological performance is maintained (Seebacher 2005). This latter approach is of particular interest because animals

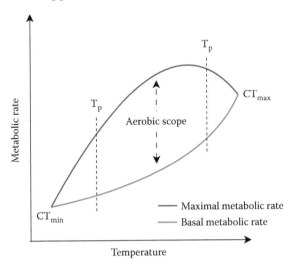

Figure 2.2 Metabolic scope. The difference between basal metabolic rate and maximal metabolic rate represents the amount of energy an animal has for performance. As animals approach their pejus temperatures (Tp), metabolic scope declines exponentially, until it collapses at the critical temperatures (CT_{min} and CT_{max})

continue to maximize fitness, that is they stay within their performance breadth, despite typically unfavorable thermal extremes. Thus, this chapter will largely focus on strategies reptiles and amphibians use to compensate for seasonal variations in temperature. First, we will provide a brief overview of recent insights into hypometabolism, followed by a review of the recent literature on mechanisms of thermal compensation in reptiles and amphibians. Our intent here is to provide an overview of the literature since Seebacher (2005), which focused exclusively on phenotypic plasticity in reptiles.

Dormancy in response to seasonal variation in temperature

Many reptiles and amphibians do not stay active year round and enter hypometabolic states when temperatures begin to approach CT_{min} or CT_{max}. Aestivation for example is a process that allows animals to undergo periods of dormancy during prolonged heat and water depletion (Young et al. 2011; Storey et al. 2012), whereas torpor and hibernation represent periods of metabolic depression associated with decreases in body temperature (Muir et al. 2013). This nomenclature is potentially confusing because in regions where the cold and dry seasons are coincident, it is never totally clear whether hypometabolism is the result of temperature or water deficit. The biochemical pathways that regulate hypometabolism are similar whether in response to extreme heat or extreme cold and are highly conserved across vertebrate groups (Storey and Storey 2007). Suppression of metabolic rate is important because animals must rely on their current energy stores for periods that can be upwards of several months (Storey and Storey 2007). Many reptiles and amphibians also overwinter in frozen rivers, ponds, and lakes, where hypometabolism is not only a strategy to conserve energy, but also a necessity to survive hypoxia (Costanzo et al. 2001; Storey 2007; Jackson and Ultsch 2010). Frogs and freshwater turtles provide the best examples of metabolic rate suppression during hibernation, where metabolic rates can be reduced by up to 99% of the typical air breathing values (Herbert and Jackson 1985; Jackson and Ultsch 2010). The exact signals for hypometabolism in these animals are not fully resolved (Bickler and Buck 2007), but there is some evidence that food deprivation may interact with cold exposure and hypoxia to signal metabolic suppression. For instance, fasting reduces respiratory chain activity in the skeletal muscles of frogs and livers of toads, but only during periods of cold exposure (Boutilier and St-Pierre 2002; Trzcionka et al. 2008). Cold exposure, hypoxia, and food deprivation may also signal metabolic suppression indirectly through combined effects on cellular energy (ratios of ATP: adenosine diphosphate [ADP]:[adenosine monophosphate] AMP) status. Energy deficiency is reflected in increased concentrations of AMP relative to ADP and ATP, which in turn activate the AMP-activated protein kinase (AMPK; Suter et al. 2006; Hardie 2008). Thus, the AMPK system can act as an energy-sensing switch and is increasingly recognized as a master controller of metabolism, particularly of glycolysis and fatty acid oxidation in skeletal muscle, liver, and heart. AMPK has therefore been suggested as a likely candidate to coordinate hypometabolism in response to unfavorable environmental conditions in ectotherms; however, its exact role in hypometabolism remains unknown (Bickler and Buck 2007; Storey et al. 2012). AMPK enhances ATP production, but it also switches off energy-consuming processes (Hardie et al. 1998; Hardie 2008) and may thereby reduce global metabolic rates. Accordingly, it has been shown that AMPK is activated during entry into hypometabolic states in different reptile and amphibian groups (see Rider et al. 2011; Storey and Storey 2012).

Many studies focus on hypometabolism as a strategy that animals use to survive suboptimal temperatures that interrupt periods of growth and mating. However, developing embryos of different reptile and amphibian species can also enter hypometabolic states to delay hatching until more favorable thermal conditions arrive (Radder and Shine 2006; Doody 2011; Rafferty and Reina 2012; Du and Shine 2015). The capacity for embryonic torpor, aestivation, and delayed hatching appears to be inversely proportional to parental care, and it is therefore much more common in reptiles and amphibians than in birds and mammals (Rafferty and Reina 2012). Whereas embryonic cold torpor typically occurs for brief periods at any stage during incubation, aestivation and delayed hatching occur only in the final stages of development (Rafferty and Reina 2012).

There are a number of ways that extreme temperatures can cause physical damage to tissues, including oxidative stress and protein denaturing at high temperatures, membrane instability at both high and low temperatures, and intracellular ice crystal formation at low temperatures (Storey and Storey 1992; Pörtner 2001; O'Brien 2011). Thus, coping strategies under extreme thermal conditions typically require defense mechanisms to preserve cell integrity. Wood frogs (*Lithobates sylvaticus*) of Alaska and the Canadian North and two species of newts (*Salamandrella keyserlingii* and *Salamandrella schrenckii*) indigenous to Asian Russia can survive freezing at temperatures as low as −16°C and −35°C, respectively (Berman et al. 1984, 2010; Larson et al. 2014). These species represent the most extreme examples of cryoprotection known in vertebrates. The physiology underlying freeze tolerance in wood frogs is relatively well understood, where ice formation is restricted to extracellular spaces to prevent damage to the intracellular environment (Costanzo et al., 2013). Concentrated levels of glucose and urea act as cryoprotectants, and urea concentrations may also double as a metabolic suppressant (Muir et al. 2010; Costanzo and Lee 2013). For further review of reptile and amphibian freeze tolerance, we direct readers to Storey and Storey (1992, 2005) and Storey (2006).

Compensation strategies for seasonal variation in temperature

There are different evolutionary pathways to maintain performance in response to temperature variation at different time scales. When fluctuations in the thermal landscape occur over an evolutionary timescale (over many generations), species and populations can adapt to these changes through natural selection for fixed phenotypes (see Chapter 1). However, when thermal environments fluctuate over short periods relative to lifespan, selective pressures favor more plastic phenotypes that buffer against intragenerational shifts in temperature (Chevin et al. 2013; Furrow and Feldman 2013). Because seasonal variation occurs within the lifetimes of most reptiles and amphibians, we restrict our discussion to plastic phenotypes that allow animals to compensate for short-term fluctuations in temperature relative to lifespan.

When body temperature is outside the performance breadth of the thermal reaction norm, there are two main changes that can be made to achieve optimal performance. Body temperature can be shifted to the optimal range of the thermal reaction norm (thermoregulation; Figure 2.3a), or the optimal range of the thermal reaction norm can shift to match current body temperature (thermal acclimation; Figure 2.3b). Reptiles and amphibians use both of these strategies to compensate for seasonal changes in temperature. Behavioral thermoregulation refers to the exploitation of different avenues of heat transfer to maintain stable body temperature relative to fluctuations in operative temperature. For instance, saltwater crocodiles (*Crocodylus porosus*) shuttle between basking in the sun on riverbanks to bathing in the cool river waters to regulate their mean body temperatures

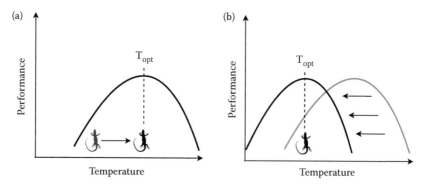

Figure 2.3 Compensatory thermal responses. Behavioral thermoregulation (a) is a process whereby animals regulate their body temperatures to the optimal temperature (T_{opt}) of their thermal reaction norm, whereas thermal acclimation (b) is a physiological conditioning whereby the thermal reaction norm is shifted to optimize performance for the new body temperature.

to 29°C in summer (Glanville and Seebacher 2006). Because animal behavior is relatively dynamic, this strategy can buffer animals from relatively rapid changes in the thermal environment. Thermal acclimation on the other hand buffers animal physiology from fluctuations in body temperature. This process is a reversible conditioning that occurs over a relatively longer period of time, usually days to weeks. During acclimation, animals remodel physiological pathways to shift their thermal reaction norms, thereby maintaining performance under otherwise limiting conditions. While the former strategy regulates body temperature to exploit thermal opportunities in the environment, the latter regulates thermal sensitivities to buffer the limitations for body temperature regulation. Despite the fundamentally different approaches, however, these two strategies are not necessarily mutually exclusive.

Here, we review recent insights into the evolutionary complexities of acclimatory responses in amphibians and reptiles and how these responses are, sometimes, integrated with behavioral changes. For a more extensive review focused primarily on behavioral thermoregulation, we refer readers to Seebacher and Franklin (2005) and Bicego et al. (2007).

Thermal acclimation

The terms acclimation and adaptation are sometimes confused in the scientific literature, and there is a tendency to separate the two into mutually exclusive categories. Acclimation is a form of phenotypic plasticity that acts at the level of the individual to buffer against environmental changes within its lifetime (Fry and Hart 1948; Guderley 1990; Wilson and Franklin 2002). Adaptation on the other hand acts at the population level to buffer environmental changes over multiple generations (Lande and Shannon 1996; Hermisson and Pennings 2005). The capacity for acclimation itself represents an evolved genetic trait that can vary among individuals, populations, and species (Seebacher et al. 2012; Hossack et al. 2013), and is therefore subject to the same selective pressures that drive adaptation.

During thermal acclimation, ectotherms remodel physiological pathways to shift their thermal reaction norms, specifically in the range of their performance breadth to maintain performance in otherwise limiting conditions. Thermal acclimation has been studied in a wide range of animals, including both vertebrates (Guderley 2004a,b; Seebacher 2009; O'Brien 2011) and invertebrates (Sokolova and Pörtner 2003; Koštál et al. 2011; Ronges et al. 2012). It is particularly important, however, in species or populations that face seasonal temperature variation larger than the daily average temperature variation (Sinclair et al. 2006). It is also particularly important in those species or populations that have limited capacities for behavioral thermoregulation (Lowe et al. 2010). This may be true of many aquatic reptiles and amphibians that inhabit waters with relatively high thermal homogeneity, or in ecological contexts where costs associated with thermoregulatory behaviors, such as predation risks, outweigh the benefits.

Acclimation of locomotor performance is a major area of study because of its importance for animal fitness. Locomotor performance has been linked to many life history traits that contribute to survival and reproduction in reptiles and amphibians, including prey capture and predator escape (Miles 2004; Higham 2007), dispersal (Seebacher and Franklin 2011), social dominance (Perry et al. 2004), and mating (Husak et al. 2006; Peterson and Husak 2006; Husak and Fox 2008). In variable climates, there would be, therefore, strong selection for individuals and populations that maintain locomotor performance despite thermodynamic limitations. The semiaquatic alpine newt (*Ichthyosaura alpestris*) can acclimate both swimming performance and running performance to seasonal changes in temperature, but the underlying regulatory mechanisms that coordinate each mode of locomotion rely on different thermal cues (Šamajová and Gvoždík 2010). The thermal sensitivity of swimming performance shifts in response to the mean acclimation temperature, whereas the thermal sensitivity of running performance is more influenced by diel variations in temperature (Šamajová and Gvoždík 2010). This work shows that acclimation responses may be more dynamic than previously assumed, and that although locomotor performance is a whole-animal measure, it is comprised of complex traits that are not necessarily regulated as a "whole." Instead, locomotion is supported at multiple levels of physiology, such as energy metabolism for ATP production and muscle performance for power output (skeletal) and oxygen supply (cardiac).

Acclimation of energy metabolism

A major challenge that reptiles and amphibians face in variable thermal environments is the maintenance of energy metabolism so that ATP production continues to meet demand (Brand 1997). Thermal acclimation of aerobic metabolism can be mediated by phenotypic changes at multiple levels of organization, including changes in enzyme concentration, mitochondrial biogenesis, modification of cellular and/or mitochondrial membranes, conformational changes, and isoform profiles that optimize enzyme efficiencies at different temperatures (see Seebacher 2005; Schulte 2015). In the last 15 years, there has been a paradigm shift whereby changes in gene expression, rather than gene sequence, have been recognized as the principal factors determining phenotype (Gerstein et al. 2007). Thus, knowing how gene expression is regulated is fundamental to understanding animal function.

It is becoming increasingly evident that the transcriptional programs that coordinate metabolism during thermal acclimation in reptiles and amphibians are remarkably homologous with other vertebrates.

For instance, the coactivator peroxisome proliferator-activated receptor gamma coactivator 1 alpha (PGC-1α) regulates metabolic heat production in birds and mammals (Puigserver et al. 1998; Wu et al. 1999; Puigserver and Spiegelman 2003; Liang and Ward 2006) and is also involved in cold acclimation in fish (LeMoine et al. 2008; Little et al. 2013). Evidently, PGC-1α also responds to cold acclimation in reptiles and amphibians, in which it is upregulated in the livers of saltwater crocodiles (*C. porosus*; Seebacher et al. 2009) and the skeletal muscle of European common lizards (*Lacerta vivipara*; Rey et al. 2008). In mammals, PGC-1α binds transcription factors to upregulate the expression of many metabolic genes (Scarpulla 2011). Specifically, PGC-1α regulates metabolic phenotypes by interacting with the two classes of transcription factors, the peroxisome proliferator-activated receptors and the nuclear respiratory factors (NRF-1 and NRF-2) (Puigserver et al. 1998; Scarpulla 2008, 2011). The downstream targets of PGC-1α in cold acclimated reptiles and amphibians appear to be similar to its targets in other vertebrate groups. Enzymes that play regulatory roles in aerobic metabolism, such as citrate synthase, cytochrome oxidase, and F_0F_1 ATP-synthase, are upregulated or have increased activities during cold acclimation in different reptilian and amphibian species (Berner and Bessay 2006; Seebacher et al. 2009; Guderley and Seebacher 2011). Perhaps more surprisingly, however, is that reptilian homologs to the mammalian uncoupling protein (UCP1) are also upregulated by PGC-1α during cold acclimation (Rey et al. 2008; Schwartz et al. 2008). The mammalian UCP1 promotes thermogenesis by uncoupling oxidative phosphorylation (Klaus et al. 1991), but the physiological role(s) of its homologs in vertebrate ecto-therms are not currently known. However, they have been suggested to enhance the efficiency of aerobic metabolism at low temperatures (Guderley and Seebacher 2011) and limit production of reactive oxygen species (Rey et al. 2008) during cold acclimation. While these predictions remain to be tested, the overwhelming similarity in response to chronic cold exposure between reptiles and amphibians and other vertebrate groups suggests that different thermoregulatory strategies may share a common evolutionary basis (Little and Seebacher 2014).

Acclimation of muscle function
Skeletal muscle supplies the power required for locomotion, where muscle function depends on the contractile properties of the muscle (myofibrils) and calcium handling pathways to regulate contraction and relaxation (Berchtold et al. 2000; Capitanio et al. 2006; Seebacher et al. 2012). In addition to regulating the metabolic pathways that produce energy for muscle function, the saltwater crocodile (*Crocodylus prosus*) can also adjust biomechanical properties of muscle to maintain performance at colder temperatures (Seebacher and James 2008). While isometric stress (force per cross-sectional area) of the main swimming muscle (caudofemoralis)

did not change with acclimation temperature, force development and relaxation times were significantly shorter in cold acclimated crocodiles (Seebacher and James 2008). Increased rates of force development and relaxation should promote quicker contraction–relaxation cycles, and therefore contribute to faster stride frequencies in cold conditions (Seebacher and James 2008). Cold acclimated crocodiles also produce more power output relative to their warm acclimated counterparts, which is at least partially explained by their increased rates of force development and relaxation times. However, there is no evidence that increased rates of myofibrillar ATPase activity are responsible for the increased rates of force development during cold acclimation (Seebacher and James 2008). Upon excitation, Ca^{2+} is released from the sarcoplasmic reticulum via ryanodine receptor. The binding of Ca^{2+} activates myofibrillar ATPase to produce a power-stroke. Following the power stroke, the sarco-endoplasmic reticulum Ca^{2+}-ATPase (SERCA) sequesters calcium back into the sarcoplasmic reticulum to replenish calcium stores for subsequent contractions (Periasamy and Kalyanasundaram 2007). Thus, SERCA represents a likely mechanism to modulate rates of contraction frequency during cold acclimation, as is suggested to be the case in zebrafish (Little and Seebacher 2013).

Recent work in African clawed frogs (*Xenopus laevis*) and cane toads (*B. marinus*) has shown that locomotor performance becomes more costly at low temperatures, which is likely to be the result of concurrent increases in muscle viscosity and stiffness at cold temperatures (Seebacher and Franklin 2011; Seebacher et al. 2014). In frogs, muscle oxygen consumption per unit work increases at low temperatures, regardless of acclimation treatment (Seebacher et al. 2014). Thus, in addition to the slower reaction rates of enzymes that coordinate calcium handling and force-production, cold exposure imposes inherent limitations on the muscle fiber itself, which represents another barrier to performance. Locomotor performance relies on different levels of physiology, but they do not operate in isolation. In other words, remodeling biomechanical pathways will only help compensate for the effects of cold exposure if energy metabolism is simultaneously upregulated to meet the increased costs of power output.

As in skeletal muscle, low temperatures can constrain cardiac function by compromising the pathways that coordinate contraction strength and frequency. Reduced cardiac scope (the difference between resting cardiac output and maximum cardiac output) means that the animal will have a reduced capacity to deliver oxygen to working muscles. Because the heart is critical to sustaining maintenance-level physiology, however, it presumably faces strong selection pressures to optimize function with changes in temperature. There has been relatively little work on thermal acclimation of cardiac performance, but one recent study highlights the importance of maintaining cardiac function for several reptilian embryos (Du et al. 2010). As stated earlier, thermal acclimation is particularly important in animals that have limited capacity to thermo-regulate behaviorally (Lowe et al. 2010), and embryos provide an extreme example. When cold acclimated, embryos of the common snapping turtle (*Chelydra serpentina*), the eastern fence lizard (*Sceloporus undulatus*), and the stripe-tailed rattlesnake (*Elaphe taeniura*) were all able to partially compensate their heart rates for decreases in ambient temperatures (Du et al. 2010). Partial compensation for the thermal sensitivity of heart rate would increase the relative oxygenation of tissues during embryogenesis, presumably increasing egg survival and expediting development (Du et al. 2010). In the same study, embryos of the eastern three-lined skink (*Bassiana duperreyi*) did not acclimate heart rate during cold exposure. One possible explanation is that embryos that acclimate heart rate were found in relatively deep nests where temperatures remained more or less stable, whereas embryos of the eastern three-lined skink were found in shallow nests with higher thermal variability (Du et al. 2010). Thus, the evolution-ary benefits of cold acclimation in embryos of the eastern three-lined skink may be outweighed by potential costs associated with daily rises in temperature. An important note, however, concerns whether this effect is in fact reversible acclimation or developmental plasticity. In other words, it is not clear from this study whether embryos can reversibly acclimate to shifts in environmental temperature during embryonic development, or whether early temperatures fix an embryonic phenotype that performs optimally only at that temperature. The latter strategy, developmental plasticity, is discussed later in this chapter.

Regulation of acclimation

Despite the importance of acclimation for physiology, ecology, and con-servation, the overarching pathways that coordinate this response in ectotherms remain unresolved. The same hypothalamic centers that control thermoregulatory behavior in ectotherms may also regulate the thermal acclimation response (Seebacher 2009). Sympathetic output from these regulatory centers stimulate β-adrenergic receptors on cell surfaces at target tissues, which can signal transcriptional programs involved in the acclimation response (for review see Seebacher 2009). Thyroid hor-mones have recently been shown to play a key regulatory role in ther-mal acclimation in fish. Hypothyroidism abolishes the capacity for thermal acclimation and reduces whole-animal performance (see Little and Seebacher 2014). Importantly, different target tissues have different sensitivities to circulating levels of thyroid hormones, and this sensitiv-ity depends on the thermal history of the fish. While there are no recent studies on thyroid-regulated acclimation in amphibian and reptilian spe-cies, very early studies suggest that at least some aspects of this pathway are conserved (Locker and Weish 1966; Lagerspetz et al. 1974; Packard and Packard 1975; Packard and Randall 1975). In leopard frogs (*Rana pipiens*), there are correlative effects of thermal acclimation on thyroid function and aerobic metabolism (Lagerspetz et al. 1974), and treatment with thyroid hormones mimics the effects of cold acclimation (Locker and Weish 1966). In European common frogs (*Rana temporaria*), thyroid hormones increase metabolic rates of liver tissues when the animals are acclimated to warm environments, but has no effect during cold acclimation (Packard and Packard 1975). Thus, assessment of thyroid hormones in relation to thermal acclimation, and perhaps even thermoregulatory behavior, may be a worthwhile pursuit in reptiles and amphibians.

Coevolution of thermoregulation and thermal acclimation

Behavioral thermoregulation and thermal acclimation represent two fundamentally different strategies ectotherms use to overcome variation in their thermal environments. During behavioral thermoregulation, ectotherms subsample the environment, thereby reducing the experienced environmental variability and the gap between environmental temperature and optimal body temperature. Behavioral programs are highly complex and derived from delicately regulated interacting gene networks that control neurochemical pathways (Cooper 2002; Seebacher and Murray 2007; Garrity et al. 2010). In the case of thermoregulatory behaviors, these pathways are likely to have coevolved with set-points in the brain and thermal sensitivities of performance traits (Angilletta et al. 2006; Boulant 2006). Behavioral thermoregulation therefore buffers body temperature from variation in ambient thermal conditions, whereas acclimation buffers biochemical and physiological rate functions from changes in body temperature. Despite their differences, these processes can be complementary. Many reptiles will select different body temperatures during different seasons. This represents a tradeoff whereby costs associated with thermoregulatory behavior are offset by altering preferred body temperatures (Seebacher et al. 2003; Gvoždík 2012). Saltwater crocodiles, for instance, will select significantly lower preferred body temperatures in winter conditions to reduce thermoregulatory costs (Glanville and Seebacher 2006). At the same time, however, they can offset the costs of reduced body temperatures by acclimating biological rate functions to their new preferred body temperature (Seebacher et al. 2003; Glanville and Seebacher 2006; Seebacher and James 2008; Guderley and Seebacher 2011).

Tradeoffs between thermoregulatory behavior and thermal acclimation also occur in amphibian species. The eastern red-spotted newt (*Notophthalmus viridescens viridescens*) is active year round and even breeds in ponds encrusted in ice over winter (Berner and Puckett 2010). Preferred body temperatures are plastic and depend on season, where summer-acclimated newts prefer higher body temperatures and winter-acclimated newts prefer lower body temperatures (Berner and Puckett 2010). Despite reduced body temperatures, winter-acclimated newts have higher standard metabolic rates and enzyme reaction rates relative to their summer-acclimated counterparts (Berner and Puckett 2010). This means that thermoregulatory behaviors and thermal acclimation of metabolic tissues are integrated in these animals so that preferred body temperatures shift with thermal reaction norms during acclimation.

These studies demonstrate how behavioral thermoregulation and thermal acclimation do not necessarily represent mutually exclusive thermal responses. They have coevolved in at least some species to help offset potential costs associated with each strategy. It is difficult to reconcile whether one process drives another, or both occur simultaneously. For instance, individuals may choose lower body temperatures that subsequently drive acclimatory response to that temperature. Alternatively, lower environmental temperatures may promote acclimation responses in tissues, the benefits of which are then maximized behaviorally by lowering preferred body temperatures.

Developmental plasticity

Environmental conditions during development can alter offspring phenotypes irreversibly by acting either directly on the offspring (intragenerational effects) or via signals passed on from the parents (intergenerational effects). The mechanisms underlying these effects are becoming better understood and are currently at the cutting edge of research into the link between phenotype and genotype. Parental effects can be mediated by transfer of material (e.g., mitochondria) from the maternal and even paternal gametes to the offspring cells (Mousseau and Fox 1998).

Importantly, both parents can also affect offspring phenotypes by altering DNA molecules and thereby influencing offspring gene expression (Ng et al. 2010; Richards et al. 2010). DNA molecules can be altered chemically by DNA-methyltransferases that transfer a methyl group from the S-adenosyl methionine to the cytosine ring (Klose and Bird 2006). Methylation of DNA results in gene silencing by restricting access of transcriptional regulators, or by direct suppression of methylated cytosine–guanine dinucleotides (Bird and Wolffe 1999), and it is a potent mechanism of programming gene expression (Kappeler and Meaney 2010). There are several methyltransferases, which either copy existing methylation patterns to the new DNA strand during replication for intergenerational effects (DNMT1; Klose and Bird 2006), or introduce novel cytosine methylation for intragenerational effects (DNMT3a and DMNT3b; Badyaev 2014). Histone modifications comprise another epigenetic mechanism that regulates gene expression intragenerationally (Weaver et al. 2004). During histone acetylation for instance, acetyl groups are transferred to or removed from histones by acetyltransferases and histone deacetylases, respectively (Grunstein 1997; Kouzarides 2007). Histone acetylation relaxes chromatin structure to promote access to transcriptional complexes, whereas histone deacetylation promotes histone binding to the DNA and thereby restricts the access of nearby genes (Grunstein 1997; Kouzarides 2007).

The field of epigenetics is still in its infancy, but there is promise that these mechanisms will help to explain the developmental strategies many reptiles and amphibians use to cope with temperature stress (Storey et al. 2012; Tattersall et al. 2012). Methylation levels in amphibians are significantly higher than those in mammals and birds, a trend that is suggested to result from their lower average body temperatures (Varriale 2014). However, average body temperatures typically covary with body temperature stability, where birds and mammals have relatively stable and high body temperatures and reptiles and amphibians have lower and more variable body temperatures. Higher rates of methylation in amphibians may therefore represent a developmental strategy to optimize offspring phenotype to thermal environments predicted by parental body temperatures. The little work that has been done in reptiles and amphibians focuses more on dormancy than compensatory response. For instance, histone deacetylation and dephosphorylation have been linked to transcriptional silencing during metabolic rate depression in red-eared sliders (*Trachemys scripta elegans*) and striped burrowing frogs (*Cyclorana alboguttata*; see Storey et al. 2012). Due to the relative flexibility (and reversibility) of histone modifications, it is likely that these mechanisms transcend development and play important regulatory roles in mature animals (Storey et al. 2012; Tattersall et al. 2012). Whether they also guide reversible acclimation responses, however, remains to be studied.

Thermal biology and climate change

Reptiles and amphibians are among the most sensitive groups to climate change, and both have already experienced global declines and extinctions (see Urban et al. 2014). As discussed, thermal constraints play a major role in setting geographical range limits for many species and populations (Pörtner 2001, 2002; Kearney and Porter 2009; Seebacher and Franklin 2011; McCann et al. 2014). While individuals will survive until they reach CT_{min} or CT_{max}, populations and species will only survive if enough individuals can remain within their performance breadths. There has been a trend in climate change biology to model future distributions of species and populations based solely on predicted shifts in average temperature. Therefore, populations that cannot track their current climatic niches as they shift or shrink across the landscape are assumed to face extinction. These forecasts, however, fail to account for animal responses.

Depending on the time course of thermal change relative to lifespan, adaptation, and more dynamic responses like behavioral thermoregulation, thermal acclimation and developmental plasticity will be more likely to determine population- and species-level survival. Regardless of mean temperature, thermal variation can also have strong effects on reptile and amphibian performance and should therefore be incorporated into such models (Clusella-Trullas et al. 2011; Sunday et al. 2012, 2014).

The current rate of climate change is quick relative to the average generation time of most reptiles and amphibians. Therefore, genetic adaptation across several generations is likely to be ineffective to buffer against the effects of climate change. A recent meta-analysis (Urban et al. 2014) found no evidence that adaptive evolution will offset the effects of climate change in reptiles and amphibians. In the same study, however, plastic responses within individuals were predicted to buffer populations and species against some of the more formidable threats of climate change (see Urban et al. 2014). The intent here is not to suggest that phenotypic plasticity, whether behavioral, physiological, or developmental, will leave reptile and amphibian species untouched by climate change, but rather that these are important responses that should be considered when assessing animal vulnerability to predicted climate change scenarios.

References

Angilletta, M. J. Jr., Bennett, A. F., Guderley, H., Navas, C. A., Seebacher, F., and Wilson, R. S. 2006. Coadaptation: A unifying principle in evolutionary thermal biology. *Physiol. Biochem. Zool.* 79: 282–294.

Angilletta, M. J. Jr., Huey, R.B., and Frazier, M. R. 2010. Thermodynamic effects on organismal performance: Is hotter better? *Physiol. Biochem. Zool.* 83(2): 197–206.

Badyaev, A. V. 2014. Epigenetic resolution of the 'curse of complexity' in adaptive evolution of complex traits. *J. Physiol.* 592: 2251–2260.

Berchtold, M. W., Brinkmeier, H., and Müntener, M. 2000. Calcium ion in skeletal muscle: Its crucial role for muscle function, plasticity, and disease. *Physiol. Rev.* 80: 1215–1265.

Berman, D. I., Leirikh, A. N., and Meshcheryakova, E. N. 2010. The Schrenck newt (*Salamandrella schrenckii*, Amphibia, *Caudata*, *Hynobiidae*) is the second amphibian that withstands extremely low temperatures. *Dokl. Biol. Sci.* 431: 131–134.

Berman, D. I., Leirikh, A. N., and Mikhailova, E. I. 1984. Winter hibernation of the Siberian salamander *Hynobius keyserlingi. J. Evol. Biochem. Physiol.* 3: 323–327.

Berner, N. J. and Bessay, E. P. 2006. Correlation of seasonal acclimatization in metabolic enzyme activity with preferred body temperature in the eastern red spotted newt (*Notophthalmus viridescens viridescens*). *Comp. Biochem. Physiol. A Mol. Integrat. Physiol.* 144: 429–436.

Berner, N. J. and Puckett, R. E. 2010. Phenotypic flexibility and thermoregulatory behavior in the eastern red-spotted newt (*Notophthalmus viridescens viridescens*). *J. Exp. Zool. A Ecol. Genet. Physiol.* 313: 231–239.

Bicego, K. C., Barros, R. C., and Branco, L. G. 2007. Physiology of temperature regulation: Comparative aspects. *Comp. Biochem. Physiol. A Mol. Integrat. Physiol.* 147: 616–639.

Bickler, P. E. and Buck, L. T. 2007. Hypoxia tolerance in reptiles, amphibians, and fishes: Life with variable oxygen availability. *Annu. Rev. Physiol.* 69: 145–170.

Bird, A. P. and Wolffe, A. P. 1999. Methylation-induced repression-belts, braces, and chromatin. *Cell* 99: 451–454.

Boulant, J. A. 2006. Neuronal basis of Hammel's model for set-point thermoregulation. *J. Appl. Physiol.* 100: 1347–1354.

Boutilier, R. G., Donohoe, P. H., Tattersall, G. J., and West, T. G. 1997. Hypometabolic homeostasis in overwintering aquatic amphibians. *J. Exp. Biol.* 200: 387–400.

Boutilier, R. G. and St-Pierre, J. 2002. Adaptive plasticity of skeletal muscle energetics in hibernating frogs: Mitochondrial proton leak during metabolic depression. *J. Exp. Biol.* 205: 2287–2296.

Brand, M. D. 1997. Regulation analysis of energy metabolism. *J. Exp. Biol.* 200: 193–202.

Capitanio, M., Canepari, M., Cacciafesta, P. et al. 2006. Two independent mechanical events in the interaction cycle of skeletal muscle myosin with actin. *Proc. Natl. Acad. Sci.* 103: 87–92.

Chevin, L. M., Collins, S., and Lefèvre, F. 2013. Phenotypic plasticity and evolutionary demographic responses to climate change: Taking theory out to the field. *Funct. Ecol.* 27: 967–979.

Clusella-Trullas, S., Blackburn, T. M., and Chown, S. L. 2011. Climatic predictors of temperature performance curve parameters in ectotherms imply complex responses to climate change. *Am. Nat.* 177: 738–751.

Cooper, K. E. 2002. Some historical perspectives on thermoregulation. *J. Appl. Physiol.* 92: 1717–1724.

Costanzo, J. P. Jones, E. E., and Lee, R. E. Jr. 2001. Physiological responses to supercooling and hypoxia in the hatchling painted turtle, *Chrysemys picta*. *J. Comp. Physiol. B* 171: 335–340.

Costanzo, J. P., do Amaral, M. C. F., Rosendale, A. J., and Lee, R. E. 2013. Hibernation physiology, freezing adaptation and extreme freeze tolerance in a northern population of the wood frog. *J. Exp. Biol.* 216: 3461–3473.

Costanzo, J. P. and Lee, R. E. 2013. Avoidance and tolerance of freezing in ectothermic vertebrates. *J. Exp. Biol.* 216: 1961–1967.

Doody, J. S. 2011. Environmentally cued hatching in reptiles. *Integr. Comp. Biol.* 51: 49–61.

Du, W. G. and Shine, R. 2015. The behavioural and physiological strategies of bird and reptile embryos in response to unpredictable variation in nest temperature. *Biol. Rev.* 90: 19–30.

Du, W. G., Warner, D. A., Langkilde, T., Robbins, T., and Shine, R. 2010. The physiological basis of geographic variation in rates of embryonic development within a widespread lizard species. *Am. Nat.* 176: 522–528.

Fry, F. A. and Hart, J. S. 1948. Cruising speed of goldfish in relation to water temperature. *J. Fish. Res. Board Can.* 7: 169–175.

Furrow, R. E. and Feldman, M. W. 2013. Genetic variation and the evolution of epigenetic regulation. *Evolution* 68: 673–683.

Garrity, P. A., Goodman, M. B., Samuel, A. D., and Sengupta, P. 2010. Running hot and cold: Behavioral strategies, neural circuits, and the molecular machinery for thermotaxis in *C. elegans* and drosophila. *Genes Dev.* 24: 2365–2382.

Gerstein, M. B., Bruce, C., Rozowsky J. S. et al. 2007. What is a gene, post-ENCODE? History and updated definition. *Genome Res.* 17: 669–681.

Glanville, E. J. and Seebacher, F. 2006. Compensation for environmental change by complementary shifts of thermal sensitivity and thermoregulatory behaviour in an ectotherm. *J. Exp. Biol.* 209: 4869–4877.

Grunstein, M. 1997. Histone acetylation in chromatin structure and transcription. *Nature* 389: 349–352.

Guderley, H. 1990. Functional significance of metabolic responses to thermal acclimation in fish muscle. *Am. J. Physiol. Reg. I.* 259: R245–R252.

Guderley, H. 2004a. Metabolic responses to low temperature in fish muscle. *Biol. Rev.* 79: 409–427.

Guderley, H. 2004b. Locomotor performance and muscle metabolic capacities: Impact of temperature and energetic status. *Comp. Biochem. Physiol. B Comp. Biochem.* 139: 371–382.

Guderley, H. and Seebacher, F. 2011. Thermal acclimation, mitochondrial capacities and organ metabolic profiles in a reptile (*Alligator mississippiensis*). *J. Comp. Physiol. B* 181: 53–64.

Gvoždík, L. 2012. Plasticity of preferred body temperatures as means of coping with climate change? *Biol. Lett.* 8: 262–265.

Hardie, D. G. 2008. AMPK: A key regulator of energy balance in the single cell and the whole organism. *Int. J. Obesity* 32: S7–S12.

Hardie, D. G., Carling, D., and Carlson, M. 1998. The AMP-activated/SNF1 protein kinase subfamily: Metabolic sensors of the eukaryotic cell? *Annu. Rev. Biochem.* 67: 821–855.

Hazel, J. R. 1995. Thermal adaptation in biological membranes: Is homeoviscous adaptation the explanation? *Annu. Rev. Physiol.* 57: 19–42.

Herbert, C. V. and Jackson, D. C. 1985. Temperature effects on the responses to prolonged submergence in the turtle Chrysemys picta bellii. II. Metabolic rate, blood acid-base and ionic changes, and cardiovascular function in aerated and anoxic water. *Physiol. Zool.* 670–681.

Hermisson, J. and Pennings, P. S. 2005. Soft sweeps molecular population genetics of adaptation from standing genetic variation. *Genetics* 169: 2335–2352.

Higham, T. E. 2007. The integration of locomotion and prey capture in vertebrates: Morphology, behavior, and performance. *Integr. Comp. Biol.* 47: 82–95.

Hossack, B. R., Lowe, W. H., Webb, M. A., Talbott, M. J., Kappenman, K. M., and Corn, P. S. 2013. Population-level thermal performance of a cold-water ectotherm is linked to ontogeny and local environmental heterogeneity. *Freshwater Biol.* 58: 2215–2225.

Huey, R. B., Kearney, M. R., Krockenberger, A., Holtum, J. A., Jess, M., and Williams, S. E. 2012. Predicting organismal vulnerability to climate warming: Roles of behaviour, physiology and adaptation. *Philos. Trans. R. Soc. Lond. B Biol. Sci.* 367: 1665–1679.

Huey, R. B. and Kingsolver, J. G. 1989. Evolution of thermal sensitivity of ectotherm performance. *Trends Ecol. Evol.* 4: 131–135.

Huey, R. B. and Stevenson, R. D. 1979. Integrating thermal physiology and ecology of ectotherms: A discussion of approaches. *Am. Zool.* 19: 357–366.

Husak, J. F. and Fox, S. F. 2008. Sexual selection on locomotor performance. *Evol. Ecol. Res.* 10: 213.

Husak, J. F., Fox, S. F., Lovern, M. B., and Bussche, R. A. 2006. Faster lizards sire more offspring: Sexual selection on whole-animal performance. *Evolution* 60: 2122–2130.

Jackson, D. C. and Ultsch, G. R. 2010. Physiology of hibernation under the ice by turtles and frogs. *J. Exp. Zool. A Ecol. Genet. Physiol.* 313: 311–327.

Kappeler, L. and Meaney, M. J. 2010. Epigenetics and parental effects. *Bioessays* 32: 818–827.

Kearney, M. and Porter, W. 2009. Mechanistic niche modelling: Combining physiological and spatial data to predict species' ranges. *Ecol. Lett.* 12: 334–350.

Klaus, S., Casteilla, L., Bouillaud, F., and Ricquier, D. 1991. The uncoupling protein UCP: A membraneous mitochondrial ion carrier exclusively expressed in brown adipose tissue. *Int. J. Biochem.* 23: 791–801.

Klose, R. J. and Bird, A. P. 2006. Genomic DNA methylation: The mark and its mediators. *Trends Biochem. Sci.* 31: 89–97.

Koštál, V., Korbelová, J., Rozsypal, J. et al. 2011. Long-term cold acclimation extends survival time at 0°C and modifies the metabolomic profiles of the larvae of the fruit fly *Drosophila melanogaster*. *PLoS One* 6: e25025.

Kouzarides, T. 2007. Chromatin modifications and their function. *Cell* 128: 693–705.

Lagerspetz, K. Y. H., Harri, M. N. E., and Okslahti, R. 1974. The role of the thyroid in the temperature acclimation of the oxidative metabolism in the frog *Rana temporaria*. *Gen. Comp. Endocrinol.* 22: 169–176.

Lande, R. and Shannon, S. 1996. The role of genetic variation in adaptation and population persistence in a changing environment. *Evolution* 50: 434–437.

Larson, D. J., Middle, L., Vu, H., Zhang, W., Serianni, A. S., Duman, J., and Barnes, B. M. 2014. Wood frog adaptations to overwintering in Alaska: New limits to freezing tolerance. *J. Exp. Biol.* 217: 2193–2200.

LeMoine, C. M., Genge, C. E., and Moyes, C. D. 2008. Role of the PGC-1 family in the metabolic adaptation of goldfish to diet and temperature. *J. Exp. Biol.* 211: 1448–1455.

Liang, H. and Ward, W. F. 2006. PGC-1α: A key regulator of energy metabolism. *Adv. Physiol. Educ.* 30: 145–151.

Little, A. G., Kunisue, T., Kannan, K., and Seebacher, F. 2013. Thyroid hormone actions are temperature-specific and regulate thermal acclimation in zebrafish (*Danio rerio*). *BMC Biol.* 11: 26.

Little, A. G. and Seebacher, F. 2013. Thyroid hormone regulates muscle function during cold acclimation in zebrafish (*Danio rerio*). *J. Exp. Biol.* 216: 3514–3521.

Little, A. G. and Seebacher, F. 2014. The evolution of endothermy is explained by thyroid hormone-mediated responses to cold in early vertebrates. *J. Exp. Biol.* 217: 1642–1648.

Locker, A. and Weish, P. 1966. Quantitative aspects of cold-adaptation and its thyroxine model in cold- and warm-blooded animals. *Helgolander Wiss. Meeresunters.* 14: 503–513.

Lowe, K., FitzGibbon, S., Seebacher, F., and Wilson, R. S. 2010. Physiological and behavioural responses to seasonal changes in environmental temperature in the Australian spiny crayfish *Euastacus sulcatus*. *J. Comp. Physiol. B.* 180: 653–660.

McCann, S., Greenlees, M. J., Newell, D., and Shine, R. 2014. Rapid acclimation to cold allows the cane toad to invade montane areas within its Australian range. *Funct. Ecol.* 28: 1166–1174.

Miles, D. B. 2004. The race goes to the swift: Fitness consequences of variation in sprint performance in juvenile lizards. *Evol. Ecol. Res.* 6: 63–75.

Mousseau, T. A. and Fox, C. W. 1998. The adaptive significance of maternal effects. *Trends Ecol. Evol.* 13: 403–407.

Muir, T. J., Costanzo, J. P., and Lee, R. E. 2010. Evidence for urea-induced hypometabolism in isolated organs of dormant ectotherms. *J. Exp. Zool. A.* 313: 28–34.

Muir, T. J., Dishong, B. D., Lee, R. E., and Costanzo, J. P. 2013. Energy use and management of energy reserves in hatchling turtles (*Chrysemys picta*) exposed to variable winter conditions. *J. Therm. Biol.* 38: 324–330.

Navas, C. A. and Carvalho, J. E. 2010. *Aestivation: Molecular and Physiological Aspects* (Vol. 49). Heidelberg: Springer-Verlag.

Ng, S. F., Lin, R. C., Laybutt, D. R., Barres, R., Owens, J. A., and Morris, M. J. 2010. Chronic high-fat diet in fathers programs [bgr]-cell dysfunction in female rat offspring. *Nature* 467: 963–966.

O'Brien, K. M. 2011. Mitochondrial biogenesis in cold-bodied fishes. *J. Exp. Biol.* 214: 275–285.

Packard, G. C. and Packard, M. J. 1975. The influence of acclimation temperature on the metabolic response of frog tissue to thyroxine administered *in vivo*. *Gen. Comp. Endocrinol.* 27: 162–168.

Packard, G. C. and Randall, J. G. 1975. The influence of thyroxine and acclimation temperature on glycogen reserves of the frog *Rana pipiens*. *J. Exp. Zool.* 191: 365–369.

Periasamy, M. and Kalyanasundaram, A. 2007. SERCA pump isoforms: Their role in calcium transport and disease. *Muscle Nerve* 35: 430–442.

Perry, G., LeVering, K., Girard, I., and Garland, T. Jr. 2004. Locomotor performance and social dominance in male *Anolis cristatellus*. *Anim. Behav.* 67: 37–47.

Peterson, C. C. and Husak, J. F. 2006. Locomotor performance and sexual selection: Individual variation in sprint speed of collared lizards (*Crotaphytus collaris*). *Copeia*, 2006: 216–224.

Pörtner, H. 2001. Climate change and temperature-dependent biogeography: Oxygen limitation of thermal tolerance in animals. *Naturwissenschaften* 88: 137–146.

Pörtner, H. O. 2002. Climate variations and the physiological basis of temperature dependent biogeography: Systemic to molecular hierarchy of thermal tolerance in animals. *Comp. Biochem. Physiol. A Mol. Integrat. Physiol.* 132: 739–761.

Pörtner, H. O. and Knust, R. 2007. Climate change affects marine fishes through the oxygen limitation of thermal tolerance. *Science* 315: 95–97.

Puigserver, P. and Spiegelman, B. M. 2003. Peroxisome proliferator-activated receptor-γ coactivator 1α (PGC-1α): Transcriptional coactivator and metabolic regulator. *Endocr. Rev.* 24: 78–90.

Puigserver, P., Wu, Z., Park, C. W., Graves, R., Wright, M., and Spiegelman, B. M. 1998. A cold-inducible coactivator of nuclear receptors linked to adaptive thermogenesis. *Cell* 92: 829–839.

Radder, R. and Shine, R. 2006. Thermally induced torpor in fullterm lizard embryos synchronizes hatching with ambient conditions. *Biol. Lett.* 2: 415–416.

Rafferty, A. R. and Reina, R. D. 2012. Arrested embryonic development: A review of strategies to delay hatching in egg-laying reptiles. *Proc. R. Soc. Lond. Biol.* 279: 2299–2308.

Rey, B., Sibille, B., Romestaing, C. et al. 2008. Reptilian uncoupling protein: Functionality and expression in sub-zero temperatures. *J. Exp. Biol.* 211: 1456–1462.

Richards, C. L., Bossdorf, O., and Pigliucci, M. 2010. What role does heritable epigenetic variation play in phenotypic evolution? *Bioscience* 60: 232–237.

Rider, M. H., Hussain, N., Dilworth, S. M., Storey, J. M., and Storey, K. B. 2011. AMP-activated protein kinase and metabolic regulation in cold-hardy insects. *J. Insect Physiol.* 57: 1453–1462.

Ronges, D., Walsh, J. P., Sinclair, B. J., and Stillman, J. H. 2012. Changes in extreme cold tolerance, membrane composition and cardiac transcriptome during the first day of thermal acclimation in the porcelain crab *Petrolisthes cinctipes*. *J. Exp. Biol.* 215: 1824–1836.

Šamajová, P. and Gvoždík, L. 2010. Inaccurate or disparate temperature cues? Seasonal acclimation of terrestrial and aquatic locomotor capacity in newts. *Funct. Ecol.* 24: 1023–1030.

Scarpulla, R. C. 2008. Transcriptional paradigms in mammalian mitochondrial biogenesis and function. *Physiol. Rev.* 88: 611.

Scarpulla, R. C. 2011. Metabolic control of mitochondrial biogenesis through the PGC-1 family regulatory network. *BBA Mol. Cell Res.* 1813: 1269–1278.

Schulte, P. M. 2015. The effects of temperature on aerobic metabolism: Towards a mechanistic understanding of the responses of ectotherms to a changing environment. *J. Exp. Biol.* 218: 1856–1866.

Schwartz, T. S., Murray, S., and Seebacher, F. 2008. Novel reptilian uncoupling proteins: Molecular evolution and gene expression during cold acclimation. *Proc. R. Soc. Lond. Biol.* 275: 979–985.

Seebacher, F. 2005. A review of thermoregulation and physiological performance in reptiles: What is the role of phenotypic flexibility? *J. Comp. Physiol. B* 175: 453–461.

Seebacher, F. 2009. Responses to temperature variation: Integration of thermoregulation and metabolism in vertebrates. *J. Exp. Biol.* 212: 2885–2891.

Seebacher, F. and Franklin, C. E. 2005. Physiological mechanisms of thermoregulation in reptiles: A review. *J. Comp. Physiol. B* 175: 533–541.

Seebacher, F. and Franklin, C. E. 2011. Physiology of invasion: Cane toads are constrained by thermal effects on physiological mechanisms that support locomotor performance. *J. Exp. Biol.* 214: 1437–1444.

Seebacher, F., Guderley, H., Elsey, R. M., and Trosclair, P. L. 2003. Seasonal acclimatisation of muscle metabolic enzymes in a reptile (*Alligator mississippiensis*). *J. Exp. Biol.* 206: 1193–1200.

Seebacher, F., Holmes, S., Roosen, N. J., Nouvian, M., Wilson, R. S., and Ward, A. J. 2012. Capacity for thermal acclimation differs between populations and phylogenetic lineages within a species. *Funct. Ecol.* 26: 1418–1428.

Seebacher, F. and James, R. S. 2008. Plasticity of muscle function in a thermoregulating ectotherm (*Crocodylus porosus*): Biomechanics and metabolism. *Am. J. Physiol. Reg. I.* 294: R1024–R1032.

Seebacher, F. and Murray, S. A. 2007. Transient receptor potential ion channels control thermoregulatory behaviour in reptiles. *PLoS One* 2: e281.

Seebacher, F., Murray, S. A., and Else, P. L. 2009. Thermal acclimation and regulation of metabolism in a reptile (*Crocodylus porosus*): The importance of transcriptional mechanisms and membrane composition. *Physiol. Biochem. Zool* 82: 766–775.

Seebacher, F., Pollard, S. R., and James, R. S. 2012. How well do muscle biomechanics predict whole-animal locomotor performance? The role of Ca^{2+} handling. *J. Exp. Biol.* 215: 1847–1853.

Seebacher, F., Tallis, J. A., and James, R. S. 2014. The cost of muscle power production: Muscle oxygen consumption per unit work increases at low temperatures in *Xenopus laevis*. *J. Exp. Biol.* 217: 1940–1945.

Sinclair, E. L., Thompson, M. B., and Seebacher, F. 2006. Phenotypic flexibility in the metabolic response of the limpet *Cellana tramoserica* to thermally different microhabitats. *J. Exp. Mar. Biol Ecol.* 335: 131–141.

Sokolova, I. M. and Pörtner, H. O. 2003. Metabolic plasticity and critical temperatures for aerobic scope in a eurythermal marine invertebrate (*Littorina saxatilis*, Gastropoda: Littorinidae) from different latitudes. *J. Exp. Biol.* 206: 195–207.

Storey, J. M. and Storey, K. B. 2005. Cold hardiness and freeze tolerance. In: *Functional Metabolism: Regulation and Adaptation.* Hoboken, NJ, USA: John Wiley and Sons, Inc., pp. 473–503.

Storey, K. B. 2006. Reptile freeze tolerance: Metabolism and gene expression. *Cryobiology* 52: 1–16.

Storey, K. B. 2007. Anoxia tolerance in turtles: Metabolic regulation and gene expression. *Comp. Biochem. Physiol. A Mol. Integrat. Physiol.* 147: 263–276.

Storey, K. B. and Storey, J. M. 1992. Natural freeze tolerance in ectothermic vertebrates. *Annu. Rev. Physiol.* 54: 619–637.

Storey, K. B. and Storey, J. M. 2004. Metabolic rate depression in animals: Transcriptional and translational controls. *Biol. Rev.* 79: 207–233.

Storey, K. B. and Storey, J. M. 2012. Aestivation: Signaling and hypometabolism. *J. Exp. Biol.* 215: 1425–1433.

Storey, K. B. and Storey, J. M. 2007. Tribute to P.L. Lutz: Putting life on 'pause' – molecular regulation of hypometabolism. *J. Exp. Biol.* 210: 1700–1714.

Storey, K. B., Storey, J. M., and Tanino, K. K. 2012. Strategies of molecular adaptation to climate change: The challenges for amphibians and reptiles. In Storey, K. B. and Tanino, K. K. (Eds.). *Temperature Adaptation in a Changing Climate: Nature at Risk*, Cambridge, MA, USA: CABI, Vol. 3, p. 98.

Sunday, J. M., Bates, A. E., and Dulvy, N. K. 2012. Thermal tolerance and the global redistribution of animals. *Nat. Clim. Change* 2: 686–690.

Sunday, J. M., Bates, A. E., Kearney, M. R., Colwell, R. K., Dulvy, N. K., Longino, J. T., and Huey, R. B. 2014. Thermal-safety margins and the necessity of thermoregulatory behavior across latitude and elevation. *PNAS* 111: 5610–5615.

Suter, M., Riek, U., Tuerk, R., Schlattner, U., Wallimann, T., and Neumann, D. 2006. Dissecting the role of 5′-AMP for allosteric stimulation, activation, and deactivation of AMP-activated protein kinase. *J. Biol. Chem.* 281: 32207–32216.

Tattersall, G. J., Sinclair, B. J., Withers, P. C. et al. 2012. Coping with thermal challenges: Physiological adaptations to environmental temperatures. *Compr. Physiol.* 2: 2151–2202.

Trzcionka, M., Withers, K. W., Klingenspor, M., and Jastroch, M. 2008. The effects of fasting and cold exposure on metabolic rate and mitochondrial proton leak in liver and skeletal muscle of an amphibian, the cane toad *Bufo marinus*. *J. Exp. Biol.* 211: 1911–1918.

Urban, M. C., Richardson, J. L., and Freidenfelds, N. A. 2014. Plasticity and genetic adaptation mediate amphibian and reptile responses to climate change. *Evol. Appl.* 7: 88–103.

Varriale, A. 2014. DNA methylation, epigenetics, and evolution in vertebrates: Facts and challenges. *Int. J. Evol. Biol.* 2014, Article number: 475981.

Weaver, I. C., Cervoni, N., Champagne, F. A. et al. 2004. Epigenetic programming by maternal behavior. *Nat. Neurosci.* 7: 847–854.

Wilson, R. S. and Franklin, C. E. 2002. Testing the beneficial acclimation hypothesis. *Trends Ecol. Evol.* 17: 66–70.

Wu, Z., Puigserver, P., Andersson, U. et al. 1999. Mechanisms controlling mitochondrial biogenesis and respiration through the thermogenic coactivator PGC-1. *Cell* 98: 115–124.

Young, K. M., Cramp, R. L., White, C. R., and Franklin, C. E. 2011. Influence of elevated temperature on metabolism during aestivation: Implications for muscle disuse atrophy. *J. Exp. Biol.* 214: 3782–3789.

chapter three

Physiological and biochemical correlates of calling behavior in anurans with different calling strategies

Catherine R. Bevier

Contents

Introduction

Male anuran amphibians rely on a variety of mating strategies and channels of communication during the breeding season to compete with other males for access to potential mates and to attract females. For example, chemical signals are used in *Litoria splendida* (Wabnitz et al. 2000), and play an important role in earless frogs that do not use bioacoustics in social communication, such as *Leiopelma hamiltoni* (Waldman and Bishop 2004). Indeed, males of species from several families of anurans, including Pipidae, Myobatrachidae, Hylidae, and Mantellidae, produce chemical signals that influence female choice (see review by Woodley [2015]). Communication using vibrational cues is exhibited by several species as they respond to prey or conspecific rivals and potential mates. Male wood frogs, *Lithobates* (*Rana*) *sylvaticus*, for example, respond with positive taxis to surface waves generated by calling and locomotion movements by conspecifics (Höbel and Kolodziej 2013). Visual signals are also effective, especially in noisy environments where acoustic signals could be masked. *Hylodes asper* (Haddad and Giarretta 1999) and *Atelopus zeteki* (Lindquist and Hetherington 1998) both breed near running water and use semaphore postures to communicate with conspecifics. Multimodal signaling has been identified for several species, where visual signals are coupled with acoustic signals as in *Epipedobates femoralis* (Narins et al. 2005), *Physalaemus pustulosus* (Taylor et al. 2008), and *Staurois parvus* (Grafe et al. 2012; review by Starnberger et al. 2013). Acoustic signals, however, are the best studied of the channels of communication in frogs, and scientists have been exploring fundamental aspects of call production for over 50 years. This chapter focuses on the great variation in call production on instantaneous, nightly, and seasonal levels and the physiological and biochemical correlates that underlie this variation.

Male frogs produce calls in different social conditions, and a recent review paper by Toledo et al. (2015) suggests the categories of reproductive, aggressive, and defensive calls. These categories include the more specific classifications recognized by Bogert (1960) and Wells (1977a,b, 1988). This chapter focuses the discussion on reproductive calls produced as advertisement calls (*sensu* Wells 1977b) directed primarily to females as potential mates. These calls are also perceived by conspecific males who may respond when locating a chorus or when competing for calling space in an established chorus.

Several layers of selection act on the production of these calls and integrate both intrinsic and extrinsic factors such as the physical environment, social environment, and physiological state of the individual. Recent work on the acoustic adaptation hypothesis (AAH) proposes that the basic call structure for many species has been shaped by selection based on habitat structure (Ziegler et al. 2011) and that individuals of some species exhibit acoustic emissions that match habitat characters to minimize degradation. For example, two subspecies of the cricket frog, *Acris crepitans* (Hylidae), are found in different habitats and produce calls that resist degradation in their respective ranges (Ryan et al. 1990). *Acris crepitans crepitans* inhabits pinewoods in eastern Texas while *Acris crepitans blanchardi* is found in more open grassland and plains habitats in western Texas, and their calls are distinct in temporal and spectral properties (Nevo and Capranica 1985; Ryan and Wilczynski 1988). As predicted from the AAH, the habitats differentially affect transmission efficiency, with less signal degradation in open habitat compared to extreme degradation in the forest. Importantly, calls of *A. c. crepitans* degraded far less in forests than those of *A. c. blanchardi*, which supports the idea that environmental selection has played a role in call divergence between the two subspecies (Ryan et al. 1990).

Frogs are rarely in single-species choruses, however, so studying the relationship between call structure and environmental features must also include aspects of how calling males partition the acoustic niche. Several studies on multispecies frog assemblages that validate the concept of niche separation among syntopic species during the breeding season have been conducted. Hödl (1977), for example, describes a community of 15 different frog species calling concurrently at a single site near Manaus, Brazil where each species exhibits a distinct advertisement call and overall calling behavior. Likewise, Martins et al. (2006) provide similar evidence for four syntopic neotropical treefrogs, and males in an eight-species chorus assemblage in a botanical garden in Thailand minimize acoustic interference by giving calls with different spectral and temporal characteristics and by varying their physical proximity (Garcia-Rutledge and Narins 2001). Males of closely related species that are syntopic in at least part of their geographic range and that are prone to hybridization may exhibit reproductive character displacement where their acoustic signals diverge in these areas compared to areas of allopatry (Howard 1993). Indeed, Höbel and Gerhardt (2003) report that male green treefrogs, *Hyla cinerea*, from populations sharing breeding sites with barking treefrogs, *Hyla gratiosa*, produce calls with significantly different spectral properties and call from higher perches than those from allopatric populations.

While these studies provide evidence that selective forces on call production in frogs stem from the extrinsic aspects of the physical and social environment, forces of sexual selection contribute significantly to the intrinsic biochemical and physiological aspects of call production as males compete in a noisy environment for the attention of potential mates. Calling behavior is more energetically costly than locomotion (Taigen and Wells 1985) and is arguably the most energetically expensive behavior that a frog may exhibit. Therefore, males of species for which overall call quality and production are key to reproductive success may be pushed to their physiological limits. Wells (2007), for example, compiled data for 13 species that exhibit high calling rates and reports that maximum oxygen consumption while calling is at least 10 times greater than that of resting metabolism. Likewise, the muscles involved in call production are fast oxidative glycolytic fibers with features that support high rates of contraction over long periods of time (Marsh and Taigen 1987). Fueling these contractions requires a mix of carbohydrates and lipids, which must be replenished for those species calling for several nights in a row. These features have contributed to a wide array of behavioral, physiological, and biochemical adaptations, which are detailed below.

Mechanisms and intrinsic correlates of call production

For most species of frogs, an acoustic signal starts in the lungs. The collection of trunk muscles (external and internal obliques, and abdominus rectus) contract to force air out of the lungs and over the vocal cords and larynx to generate vibrations, then into vocal sacs adjacent to the buccal cavity (see Wells 2007 for a detailed review). While the trunk muscles serve to initiate each call, the laryngeal apparatus plays an important role in facilitating the quality of sound production and modulation of the call. The vocal sac is the final structure into which air travels and, for most species, serves to capture and recycle air. The calling frog closes its nostrils so the air captured in the elastic vocal sacs gets shuttled back to the lungs and the cycle of call production is facilitated. The vocal sac also provides a broad surface area from which the sound can radiate (Dudley and Rand 1991), and recent investigations have identified the inflated vocal sac as an important visual complement to the acoustic signal (e.g., Narins et al. 2005; Taylor et al. 2007).

How many times this cycle continues depends on different aspects of a frog's biochemical and physiological capacity. On an instantaneous time-scale, the biochemical features of the muscles involved in call production, the collective trunk muscles, and the laryngeal dilator regulate the aerobic capacity for adenosine triphosphate (ATP) production and therefore the rate of contraction. This is reflected in temporal features of a call, including call or note production, pulse rate, and call duration. The activity level of the enzyme citrate synthase (CS) is often used as an indicator of the aerobic capacity of muscle fibers in these collective trunk and laryngeal muscles to produce ATP (e.g., Taigen et al. 1985; Bevier 1995; Carvalho et al. 2008). This enzyme catalyzes the rate-limiting step of the Krebs cycle in the mitochondrial matrix (Alp et al. 1976), so it is not surprising that frogs with higher call rates have both greater mitochondrial densities and higher activity levels of CS in muscles involved in call production (Bevier 1995; Ressel 1995). This relationship between call rate and trunk muscle CS activity holds when 20 species are compared from across climatic regimes and frog families (Figure 3.1). Other enzymes, such as β-hydroxyacyl-CoA dehydrogenase (HOAD), are important in aerobic metabolism in muscles and have also been used as markers for aerobic capacity or substrate metabolism. However, CS is a rate-limiting enzyme and therefore provides the strongest evidence to link call rate and muscle biochemistry.

Calling at high rates is likely a product of female preference for more energetically expensive calls, as confirmed by laboratory two-choice tests in which females exhibit such a preference (e.g., Murphy and Gerhardt 2000). These high calling rates require an efficient metabolism. The European treefrog, *Hyla arborea*, for example, has one of the highest rates of call production and there is great individual variation in metabolic oxygen consumption for males at a given call rate (Voituron et al. 2012). Interestingly, large males seem to be more efficient and expend less mass-specific energy than smaller males. This could allow males to spend more time calling in a chorus on a particular night or to spend more consecutive nights at a chorus; both would increase a male's chances of attracting a mate.

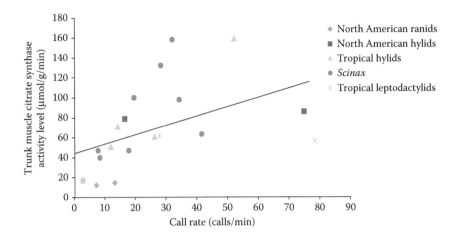

Figure 3.1 Relationship between calling rate and CS activity in the trunk muscles. Data are compiled on *A. callidryas, S. boulengeri, S. g. ruber, D. ebraccatus, D. microcephalus, T. venulosus, L. labialis,* and *P. pustulosus* from Bevier (1995), *Hypsiboas prasinus* from Kiss et al. (2009), *P. crucifer* from Taigen et al. (1985), *H. versicolor* from Wells and Taigen (1986), *L. virgatipes* from Given and McKay (1990), and *L. septentrionalis, L. clamitans, S. g. ruber, S. crospedospilus, S. hiemalis, Scinax perereca, Scinax fuscovarius,* and *Scinax rizibilis* from Bevier (unpublished data). (Linear regression, $R^2 = 0.223$, $p = 0.019$, and $y = 0.949x + 43.891$.)

Over a single period of calling lasting up to several hours, the amount of stored energy substrate in the muscles used in call production influences the overall amount of calling that can be achieved. Vertebrate animals use a mix of proteins, carbohydrates, and lipids to fuel activity and may modulate how they use these differentially for different kinds of activities (Weber 2011). Phosphocreatine is the first fuel used for muscle contractions but provides support for only a matter of seconds (Weber 2011). Carbohydrate stored in muscle tissue as glycogen is utilized next and can fuel calls over a period of hours. Within a single night of calling, frogs may use over 50% of the readily available trunk muscle glycogen stores (Wells et al. 1995; Bevier 1997b). Lipid stores provide the greatest and longest-lasting source of energy of the three and are metabolized in conjunction with carbohydrates over longer periods of time. Estimates of energy substrate utilization from investigations that generate a respiratory quotient for calling frogs ($RQ = CO_2$ released/O_2 consumed; ratios closer to 1 reflect carbohydrate oxidation while ratios closer to 0.7 reflect lipid oxidation) demonstrate that lipid stores alone could fuel up to 15 days of calling for the African reed frog (*Hyperolius marmoratus*) (Grafe 1996) and that lipids account for as much as 60% of a calling treefrog's energy expenditure (Grafe 1997). In the carpenter frog, *Lithobates virgatipes*, males are highly territorial for up to 3 months during the breeding season.

This provides an opportunity to monitor seasonal chorus attendance by individuals and to regularly take biometric data to provide information on a whole-animal level. Given (1988) reports that both body condition and lipid reserves decline remarkably over a season.

Over several days of reproductive behavior, an individual male could easily use up stored energy, but feeding regularly helps to replenish these stores and potentially augment the period over which he can call. Given the unpredictable and variable time frame over which females typically arrive at a chorus, the more time a male can spend calling, the better his chances are of attracting and mating with a female. Evidence from research on endurance rivalry and mating success in *Hyla intermedia* confirms this, where 19% of the variation in male mating success is explained by variation in the number of nights spent calling at the breeding site (Castellano et al. 2009). Furthermore, males spent more nights calling when they were in good condition at the start of the breeding season. Maintaining body condition is a challenge, but males that have opportunities to feed are likely at an advantage to maximize their time in a chorus. Results of experiments in which males are provided supplemental food during a calling period provide support. Male barking treefrogs (*H. gratiosa*) that received food supplements spent overall more time calling than unfed males (Murphy 1994), male green frogs (*Lithobates clamitans*) move less between territories (Gordon 2004), and male *H. arborea* exhibit adjustments in call rate or overall activity, depending on body size and condition, to maintain call attractiveness or overall call production (Brepson et al. 2012a).

In sum, to understand the underlying mechanisms by which male frogs may be limited in their call production at a physiological and biochemical level, exploring the source of these constraints at different temporal scales is required. How male frogs parse out their calling effort during a bout of reproductive activity provides insight into underlying mechanisms of more generalized patterns of behavior. The next section describes some of the ends of the spectra of these calling strategies, and what we understand so far.

Seasonal strategies of call production

The period over which a male frog may call in a season varies widely among species. For both temperate and tropical species, some frogs are prolonged breeders (*sensu* Wells 1977b) and call from choruses that persist for several weeks. Others call for as little as just one night out of the year and are known as explosive breeders (Wells 1977b). These two temporal patterns of breeding periods represent extremes of a spectrum, with many species exhibiting more intermediate patterns. In the neotropics, several members of the families Hylidae and Leptodactylidae exhibit prolonged breeding periods, including *Scinax boulengeri*, *Dendropsophus ebraccatus*, *Agalychnis callidryas*, and *Leptodactylus labialis*. Likewise, there are several species in which males call only for a day or two following particularly heavy rains, such as *Scinax ruber*, or exclusively at the onset of the rainy season (*Trachycephalus venulosus*).

These different calling strategies require different strategies of food acquisition and energy substrate utilization. Jönsson (1997) reviews these strategies, where the energy balance employed by using a capital breeding strategy involves food energy acquired at one point, then stored, and utilized over time. This has a storage cost but decouples feeding and reproduction to compensate for unpredictable food conditions. On the other hand, income breeders benefit from a more efficient conversion of energy intake to offspring output, but must also moderate food intake with reproductive activity.

To a degree, whether a frog species is a prolonged or explosive breeder may also be influenced by abiotic features of the environment, especially rainfall and temperature, which dictate conditions under which a frog can acquire and utilize fuel for breeding. We can draw from an optimization model from McCauley et al. (2000), developed to better understand what might drive some species to call only occasionally, and other species to call nearly every night of the season. When access to resources varies, as is the case in seasonal environments, whether a male calls or not may result from a starvation minimization strategy and an energy-state maximization strategy (McCauley et al. 2000). Males with greater energy reserves and ready access to food resources, such as male *H. gratiosa* at the start of chorus tenure and when provided food supplements (Murphy 1994), tend to call on more nights in a season and may acquire more mates. In contrast, males less likely to find food, such as *L. sylvaticus* early in the spring, may exhibit the starvation minimization strategy and compress breeding activity to a few days before retreating to foraging habitats. Some members of a breeding population may even participate as satellites, which are silent males positioned next to calling males. This latter strategy may conserve energy and is discussed in detail in a later section. In seasonally cold climates, early spring breeders provide evidence of these contrasting patterns of energy utilization. Spring peepers, *Pseudacris crucifer*, are considered a prolonged breeder and rely extensively on fat stores acquired before overwintering to fuel call production before insect prey have emerged. In contrast, woodfrogs, *L. sylvaticus*, call for only a few days and rely extensively on a short-term source of glycogen that is quickly depleted (Wells and Bevier 1997).

In many tropical areas, weather conditions seem adequate throughout a breeding season to provide good breeding conditions, but periodic and heavy rainfall seems to trigger calling activity in more explosive breeders. Two closely related hylids, *S. boulengeri* and *S. ruber*, exhibit different calling patterns that may require different patterns of energy allocation (Bevier 1997a) and support McCauley's et al. (2000) model. In brief, the more explosive breeder, *S. ruber*, maintains a higher call rate and calls for more hours per night when it is active, incurring a higher per night cost of calling than the prolonged breeder, *S. boulengeri*. This likely requires time away from the breeding site for male *S. ruber* so they can forage and sufficiently replace important energy stores. Male *S. boulengeri* have longer continuous chorus tenures, which may reflect adequate foraging between short nightly bouts of breeding. McCauley et al. (2000) note, however, that when model parameters allow for a longer breeding season and a higher probability of finding food, calling activity in *S. boulengeri* should be more episodic and interrupted by foraging activity even when calling cost was low.

Nightly variation in call production

The number of calls and the overall amount of sound that a male produces within a night varies greatly among species and among individuals within a species. For example, in tropical regions, where ambient temperature changes very little over the course of a night, frogs in the genus *Scinax* exhibit differences in call rate by an order of magnitude. Carvalho et al. (2008) and Bevier et al. (2008) describe calling behavior for several treefrogs that reflect particular patterns of call production that are well suited for each species. For example, *Scinax crospedospilus* is a prolonged breeder and very active caller, often producing calls continuously for up to 4 h in an evening (Carvalho et al. 2008). Males of this species exhibit the greatest calling effort (call rate × call duration) per hour in comparison to seven congenerics (Bevier et al. 2008). *Scinax hiemalis*, a species from the *Scinax catharinae* group, exhibits a much lower effort, 118 s/h compared to 1037 s/h for *Scinax crosspedospilus*, and calls episodically throughout a night.

Males of both species acquire food during the breeding season, but biochemical profiles of trunk muscle are quite different. The high call rate of *S. crosspedospilus* (62 calls/min) is supported by trunk muscles with high activity levels of CS to support ATP production, and relatively large lipid content (Carvalho et al. 2008). In contrast, male *S. hiemalis* call at remarkably low rates (5.7 calls/min), have trunk muscles with much lower activity levels of CS, and have lower levels of lipid in their trunk muscles (Carvalho et al. 2008).

As Bevier et al. (2008) discuss, species in this genus exhibit remarkable variation in calling behavior that are also likely influenced by local environmental conditions. *S. hiemalis* and *S. crospedospilus* are allopatric species, call during different seasons, and have very different call postures and preferred locations. Male *S. hiemalis* call episodically over a 12-h period from low perches or crevices along the banks of slow-moving water from dusk to dawn, and are typically the only frog species calling. In contrast, male *S. crospedospilus* call in a head-up position from vertical stems of ferns or other vegetation at least 1 m above the ground or water surface. Males call very actively in noisy mixed choruses at permanent ponds for up to 4 h (Bevier et al. 2008; Carvalho et al. 2008). These local conditions present very different challenges in terms of competition, and each species' behavior and physiology seem to match these conditions.

Fewer studies on individual variation in call production exist, but they do confirm the magnitude and importance of accounting for individual variation when studying relationships between calling activity and intrinsic physiological and biochemical features of the male. For example, male *P. crucifer* categorized as high-rate callers (those calling on any one night at rates greater than the chorus average over the season) are relatively larger, heavier, older, and in better body condition than low-rate calling males (Zimmitti 1999). Trunk muscles of these high-rate callers exhibit the corresponding trait of higher activity levels of CS than low-rate callers. Early in the breeding season, larger males also have significantly greater energy stores, measured as carcass lipid and glycogen, than smaller males (Duffitt and Finkler 2011). These differences are likely related to the greater energy demand required by larger males to sustain high levels of calling during the breeding season. These dramatic differences also provide a cautionary note that, where individual and seasonal differences are so varied, one must be sure to collect data and sample carefully.

Male–male competition and calling dynamics

Competition among males at a chorus can influence the duration or complexity of a call, which may also result in more energetically expensive signals. For example, male gray treefrogs, *Hyla versicolor* produce longer calls but at lower rates when in dense choruses compared to isolated males (Wells and Taigen 1986). Overall call effort was not different between these conditions, but long-calling males called for fewer hours than those males calling at higher rates with shorter calls, perhaps resulting from greater rates of energy substrate depletion. Male mink frogs, *Lithobates septentrionalis*, produce single "cuk" notes early in evening, but produce calls with more notes and with different kinds of notes during the peak period of chorus activity (Bevier et al. 2004). This strategy may be effective for competing with other males and for attracting females as they move through the chorus during the most active part of the night. The higher rate of note and call production for this species may be more energetically expensive, but no data have yet been collected to test this idea.

Males of some species may compete acoustically only with their nearest neighbors, and call interactively in duets and trios. One of these males might emerge as the leader, initiating each series of back-and-forth calls and even producing calls at higher rates than the other participants (e.g., Goin 1949; Wells and Taigen 1986; Zimmitti 1999; Bee et al. 2001). It is difficult, however, to confirm that certain individuals serve as leaders over long periods of time, that this strategy is more energetically expensive, or that call leaders are consistently more attractive to females. In experimental choruses of European treefrogs, *H. arborea*, males with the highest proportion of leading calls did have significantly higher total calling effort (call duration × call rate) but there was no relationship between call leadership and male body condition (Richardson et al. 2008). Data for male *P. crucifer* suggest that bout leaders have only slightly higher call rates than followers when males interacting in duets were observed. There were also no significant differences in trunk muscle glycogen or plasma levels of the hormone corticosterone (CORT) (Mangiamele and Bevier, unpublished data). CORT production and release is typically triggered by stress and potentially modulates calling activity (see below). More studies are needed to better elucidate whether or not leader–follower dynamics exist, if being a leader is an adaptive strategy, and whether there are physiological costs to exhibiting a leader strategy.

Neighbor interactions can persist along a different trajectory, where one male may adopt the alternative reproductive strategy of satellite behavior. Satellite males do not call but instead position themselves next to a calling male and attempt to intercept and breed with females attracted to the calling male. This strategy occurs most often in prolonged breeders and under conditions of intense competition, where reproductive success may be limited to only those males producing the most attractive calls. In many species for which males exhibit satellite behavior, the strategy is conditional, such that males may call or not on different nights and even within a night (e.g., Perrill et al. 1982; Leary et al. 2004). One hypothesis that may explain this dynamic is that a male's body condition influences whether or not he calls on a particular night. Humfeld (2013) provides evidence that satellite male green treefrogs, *H. cinerea*, began calling as soon as the associated caller was removed. But the majority of these satellites were in significantly lower body condition than their counterparts. Interestingly, these calling satellite males produced calls with temporal and spectral characters that are known to be less attractive to females. Furthermore, males in feeding experiments that did not get food produced less attractive higher frequency calls than fed males (Humfeld 2013). These different lines of evidence strongly support that temporary declines in body condition greatly influence how competitive a male can be. Data for other species, including male Woodhouse's toads, *Bufo woodhousii*, and Great Plains toads, *Bufo cognatus*, corroborate these suppositions (Leary et al. 2004). While Humfeld (2013) did not find a relationship between body size and calling strategy, smaller male *H. arborea* are more likely to be satellites than large males, though food-deprived males were not more likely to be satellites than those fed (Brepson et al. 2012b). Such mixed support for factors that may influence the onset of satellite behavior underscores the context-dependent nature of some of these conditions. Nonetheless, these reports reinforce the idea that a male's competitive success in producing more attractive calls for females may depend on his size or energetic state, especially in a dense noisy chorus and for those species that produce energetically expensive calls.

Hormones as behavioral and physiological mediators

One consequence of short- or longer-term depletion of energy reserves, a primary cause of low body condition, is the release of glucocorticoids. These are metabolic hormones that regulate carbohydrate metabolism, and CORT in particular is understood to mobilize energy stores and influence reproductive activity. In manipulated frogs, male *H. cinerea* injected with both CORT and arginine vasotocin (AVT), known to stimulate calling activity, were less likely to call than frogs injected with just AVT (Burmeister et al. 2001). This reflects an inhibitory nature of CORT on the motivation to call. It is not surprising, then, that male *H. cinerea* that adopt a satellite strategy are not only in a worse condition than their calling counterparts, but also have higher circulating levels of CORT and lower levels of testosterone (T) (Leary and Harris 2013). These levels suggest that satellite males may be physiologically stressed and at their physiological limits.

The relationships between CORT and calling behavior are more complex when androgens, such as T, are considered; CORT and T likely interact to modulate calling activity over an evening. For example, CORT and T levels were highest in male *Hyla faber* exhibiting high calling rates (Assis et al. 2012), and similar results are reported for males of the Old World frog species (Emerson and Hess 1996). Reciprocal feedback may be driving dynamics between behavior and physiology as males compete vocally for mates, so that one hormone or a behavior stimulates the release of another. Orchinik et al. (1988), for example, found high CORT levels in male *Bufo marinus* with high testosterone levels, and suggest that CORT release might be stimulated when these competitive and even agonistic males are interacting.

The energetics-hormone vocalization (EHV) model proposes that variation in hormone levels is tied with the energetic costs of vocalization, and may help to explain some of the variation in calling performance among individuals and species (Emerson 2001). As the calling activity over an evening progresses, males experience energy depletion, a subsequent increase in CORT that overrides T as a driver of calling behavior, and resulting suppression of calling behavior until a male can replenish energy reserves (Emerson 2001). Leary and Knapp (2014) offer a revised model that proposes a masking effect of stress hormones such as CORT on androgen-mediated sexually selected traits, rather than a cancelling effect. Empirical data to support the latter model is still needed, but Leary et al. (2004) tested the EHV in toads and found significantly higher CORT levels for calling male *B. cognatus* compared to noncalling males. For *Bufo woodhousei*, satellite males were smaller than calling males and also had significantly lower CORT levels. Male breeding strategies differ somewhat in these two species, but nonetheless, males alternated within or between breeding bouts with satellite and calling behavior in a way that suggests that T levels and calling activity were eventually reduced by high CORT levels. High CORT levels were associated with low T levels in satellite male *H. cinerea* (Leary and Harris 2013), which provides additional support for the hypothesis that energetic stress suppresses calling behavior. However, satellite males may be smaller or in poor body condition compared to calling rivals. Leary (2014) further explores these relationships in male *H. cinerea* and reports that, after staged close-range vocal interactions between males, contest winners successfully elicit elevated CORT levels in their rivals that effectively counter the stimulation of androgens. These contest losers subsequently take on the noncalling satellite strategy.

In general, males that exhibit satellite behavior may be energetically challenged, but also hormonally constrained from sustained calling activity, and must forage before initiating subsequent calling bouts. Emerson and Hess (2001) report that male *Pseudacris triseriata* collected early in the evening had significantly lower levels of CORT than males collected about 3 h later, and males with higher T levels also had higher CORT levels (Emerson and Hess 2001). At the species level, lower calling rates were associated with lower levels of CORT (Emerson and Hess 2001). These results support the model in that males calling at lower levels or sampled early in an evening, before the onset of energetically stressful conditions, have lower CORT levels. Calling activity, stimulated by higher T levels, can proceed without inhibition. In calling male *P. crucifer*, high levels of plasma CORT were associated with lower call rates, glycogen stores, and aerobic capacity of trunk muscles, reflected in activity levels of CS (Mangiamele and Bevier, unpublished data). This suggests that, at the time of capture, CORT levels were sufficiently high to affect calling activity. Moving forward, it will be important to pursue highly integrative studies of behavior and associated metrics of physiological capacity and hormonal regulation to better refine these relationships. For example, it is not clear over what time frame CORT begins to act on metabolism and subsequent behavioral responses, and likely depends on the immediate nutritional state of the frog. It may also be that the CORT levels of a male on any one night of calling will influence the energy available for subsequent nights of vocalization (Emerson and Hess 2001).

Conclusions

The advertisement call of male frogs is a familiar secondary sexual characteristic, and one of the most energetically expensive behaviors vertebrates produce. Different species exhibit extraordinary variation in seasonal, nightly, and hourly patterns of calling activity. Males of some species call in choruses over prolonged periods and persist for weeks or months, while others form choruses more opportunistically, for only a few days or less. Males of some species may call for several hours over a 24 h period, while other species call only briefly at night. There is a rich literature showing how these patterns of behavioral variation are influenced by local extrinsic factors of the social environment as well as intrinsic physiological and biochemical characteristics of the males and the muscles that support calling behavior. At an instantaneous time scale, female preference for a male is often based on the temporal properties of the male's call, such as call rate, and females of many species prefer males that produce more energetically expensive calls. Call rate, call intensity, and call complexity, for example, are traits influenced by the aerobic capacity of muscles used in call production, and vary dramatically both within and among species. Males able to call over longer periods of time, on either a daily or seasonal basis, have a greater potential to attract and mate with a female but may be energetically constrained. Males may also be more or less acoustically competitive with immediate neighbors, or with other males in the chorus as a whole, depending on nutritional state. These nightly and seasonal patterns of calling behavior are also influenced by levels of corticosterone produced in response to the physiological stress of this high level of activity. Interestingly, many of these correlated patterns transcend phylogenetic relationships, and these physiological characters serve as a strong selective force on behavioral capacity. While the field of behavioral energetics has moved forward with many interesting discoveries and connections, there is still much work to do to better understand the integrative nature of calling behavior in frogs.

References

Alp, P.R., E.A. Newsholme, and V.A. Zammit. 1976. Activities of citrate synthase and NAD+-linked and NADP+-linked isocitrate dehydrogenase in muscle from vertebrates and invertebrates. *Biochemical Journal* 154:689–700.

Assis, V.R., C.A. Navas, M.T. Mendonça, and F.R. Gomes. 2012. Vocal and territorial behavior in the Smith frog (*Hypsiboas faber*): Relationships with plasma levels of corticosterone and testosterone. *Comparative Biochemistry and Physiology, Part A* 163:265–271.

Bee, M.A., C.E. Kozich, K.J. Blackwell, and H.C. Gerhardt. 2001. Individual variation in advertisement calls of territorial male green frogs, *Rana clamitans*:
Implications for individual discrimination. *Ethology* 107:65–84.

Bevier, C.R. 1995. Biochemical correlates of calling activity in neotropical frogs.
Physiological Zoology 68:1118–1142.

Bevier, C.R. 1997a. Breeding activity and chorus tenure of two neotropical hylid
frogs. *Herpetologica* 53:297–311.

Bevier, C.R. 1997b. Utilization of energy substrates during calling activity in tropical frogs. *Behavioral Ecology and Sociobiology* 41:343–352.

Bevier, C.R., F.R. Gomes, and C.A. Navas. 2008. Variation in call structure and
calling behavior in treefrogs of the genus *Scinax*. *South American Journal of
Herpetology* 3:186–296.

Bevier, C.R., K. Larson, K. Reilly, and S. Tat. 2004. Vocal repertoire and calling
activity of the mink frog, *Rana septentrionalis*. *Amphibia-Reptilia* 25:255–264.

Bogert, C.M. 1960. The influence of sound on amphibians and reptiles. In Lanyon
W.E. and W.N. Tavolga (eds), *Animal Sounds and Communication*, pp. 137–320.
Washington, DC: American Institute of Biological Science.

Brepson, L., M. Troïanowski, Y. Voituron, and T. Lengagne. 2012b. Cheating for sex:
Inherent disadvantage or energetic constraint? *Animal Behavior* 84:1253–1260.

Brepson, L., Y. Voituron, and T. Lengagne. 2012a. Condition-dependent ways to
manage acoustic signals under energetic constraint in a tree frog. *Behavioral
Ecology* 24:488–496.

Burmeister, S., C. Somes, and W. Wilczynski. 2001. Behavioral and hormonal
effects of exogenous vasotocin and corticosterone in the green treefrog.
General and Comparative Endocrinology 122:189–197.

Carvalho, J.E., F.R. Gomes, and C.A. Navas. 2008. Energy substrate utilization
during nightly vocal activity in three species of *Scinax* (Anura/Hylidae).
*Journal of Comparative Physiology B: Biochemical, Systemic, and Environmental
Physiology* 178:447–456.

Castellano, S., V. Zanollo, V. Marconi, and G. Berto. 2009. The mechanisms of
sexual selection in a lek-breeding anuran, *Hyla intermedia*. *Animal Behaviour*
77:213–224.

Dudley, R. and A.S. Rand. 1991. Sound production and vocal sac inflation in the
túngara frog, *Physalaemus pustulosus* (Leptodactylidae). *Copeia* 1991:460–470.

Duffitt, A.D. and M.S. Finkler. 2011. Sex-related differences in somatic stored
energy reserves of *Pseudacris crucifer* and *Pseudacris triseriata* during the
early breeding season. *Journal of Herpetology* 45:224–229.

Emerson, S.B. 2001. Male advertisement calls: Behavioral variation and physiological processes. In Ryan, M.J. (ed.), *Anuran Communication,* pp. 36–44.
Washington, DC: Smithsonian Institution Press.

Emerson, S.B. and D.L. Hess. 1996. The role of androgens in opportunistic breeding, tropical frogs. *General and Comparative Endocrinology* 103:220–230.

Emerson, S.B. and D.L. Hess. 2001. Glucocorticoids, and rogens, testis mass, and
the energetics of vocalization in breeding male frogs. *Hormones and Behavior*
39:59–69.

Garcia-Rutledge, E.J. and P.M. Narins. 2001. Shared acoustic resources in an Old
World frog community. *Herpetologica* 57:104–116.

Given, M.F. 1988. Growth rate and the cost of calling activity in male carpenter
frogs, *Rana virgatipes*. *Behavioral Ecology and Sociobiology* 22:153–160.

Given, M.F. and D.M. McKay. 1990. Variation in citrate synthase activity in calling muscles of carpenter frogs, *Rana virgatipes*. *Copeia* 1990:863–867.

Goin, C.J. 1949. The peep order in peepers: A swamp water serenade. *Quarterly Journal of the Florida Academy of Science* 11:59–61.

Gordon, N.M. 2004. The effect of supplemental feeding on the territorial behavior of the green frog (*Rana clamitans*). *Amphibia-Reptilia* 25:55–62.

Grafe, T.U. 1996. Energetics of vocalization in the African reed frog (*Hyperolius marmoratus*). *Comparative Biochemistry and Physiology* 114A:235–243.

Grafe, T.U. 1997. Use of metabolic substrates in the gray treefrog *Hyla versicolor*: Implications for calling behavior. *Copeia* 1997:356–362.

Grafe, T.U., D. Preininger, M. Sztatecsny, R. Kasah, J.M. Dehling, S. Proksch, and W. Hödl. 2012. Multimodal communication in a noisy environment: A case study of the Bornean rock frog, *Staurois parvus*. *PLoS ONE* 7:e37965.

Haddad, C.F.B. and A.A. Giarretta. 1999. Visual and acoustic communication in the Brazilian torrent frog, *Hylodes asper* (Anura: Leptodactylidae). *Herpetologica* 55:324–333.

Höbel, G. and H.G. Gerhardt. 2003. Reproductive character displacement in the acoustic communication system of green tree frogs (*Hyla cinerea*). *Evolution* 57:894–904.

Höbel, G. and R.C. Kolodziej. 2013. Wood frogs (*Lithobates sylvaticus*) use water surface waves in their reproductive behavior. *Behaviour* 150:471–483.

Hödl, W. 1977. Call differences and calling site segregation in anuran species from Central Amazon floating meadows. *Oecologica* 28:351–363.

Howard, D.S. 1993. Reinforcement: Origin, dynamics, and fate of an evolutionary hypothesis. In Harrison, R.G. (ed.), *Hybrid Zones and the Evolutionary Process*, pp. 46–69. Oxford, UK: Oxford University Press.

Humfeld, S. 2013. Condition-dependent signaling and adoption of mating tactics in an amphibian with energetic displays. *Behavioral Ecology* 24:859–870.

Jönsson, K.I. 1997. Capital and income breeding as alternative tactics of resource use in reproduction. *Oikos* 78:57–88.

Kiss, A.C.I., J.E. de Carvalho, C.A. Navas, and F.R. Gomes. 2009. Seasonal metabolic changes in a year-round reproductively active subtropical treefrog (*Hypsiboas prasinus*). *Comparative Biochemistry and Physiology, Part A* 152:182–188.

Leary, C.J. 2014. Close-range vocal signals elicit a stress response in male green treefrogs: Resolution of an androgen-based conflict. *Animal Behaviour* 96:39–48.

Leary, C.J., T.S. Jessop, A.M. Garcia, and R. Knapp. 2004. Steroid hormone profiles and relative body condition of calling and satellite toads: Implications for proximate regulation of behavior in anurans. *Behavioral Ecology* 15: 313–320.

Leary, C.M. and S. Harris. 2013. Steroid hormone levels in calling males and males practicing alternative non-calling mating tactics in the green treefrog, *Hyla cinerea*. *Hormones and Behaviour* 64:20–24.

Leary, C.J. and R. Knapp. 2014. The stress of elaborate male traits: Integrating glucocorticoids with androgen-based models of sexual selection. *Animal Behaviour* 89:85–92.

Lindquist, E.D. and T.E. Hetherington. 1998. Semaphoring in an earless frog: The origin of a novel visual signal. *Animal Cognition* 1:83–87.

Marsh, R.L. and T.L. Taigen. 1987. Properties enhancing aerobic capacity of calling muscles in gray tree frogs *Hyla versicolor*. *American Journal of Physiology-Regulatory Integrative Comparative Physiology* 252:R786–R793.

Martins, I.A., S.C. Almeida, and J. Jim. 2006. Calling sites and acoustic partitioning in species of the *Hyla nana* and *rubicundula* groups (Anura, Hylidae). *Herpetological Journal* 16:239–247.

McCauley, S.J., S.S. Bouchard, B.J. Farina, K. Isvaran, S. Quader, D.M. Wood, and C.M. St. Mary. 2000. Energetic dynamics and anuran breeding phenology: Insights from a dynamic game. *Behavioral Ecology* 11:429–436.

Murphy, C.G. 1994. Determinants of chorus tenure in barking treefrogs (*Hyla gratiosa*). *Behavioral Ecology and Sociobiology* 34:285–294.

Murphy, C.G. and H.C. Gerhardt. 2000. Mating preference functions of individual female barking treefrogs, *Hyla gratiosa*, for two properties of male advertisement calls. *Evolution* 54:660–669.

Narins, P., W. Hodl, and D. Grabul. 2005. Bimodal signal requisite for agonistic behavior in a dart-poison frog, *Epipedobates femoralis*. *Proceedings of the National Academy of Sciences of the United States of America* 100:577.

Nevo, E. and R.R. Capranica. 1985. Evolutionary origins of ethological reproductive isolation in cricket frogs, *Acris*. *Evolutionary Biology* 19:147–214.

Orchinik, M., P. Licht, and D. Crews. 1988. Plasma steroid concentrations change in response to sexual behavior in *Bufo marinus*. *Hormones and Behavior* 22:338–350.

Perrill, S.A., H.C. Gerhardt, and R. Daniel. 1982. Mating strategy shifts in male green treefrogs (*Hyla cinerea*): An experimental study. *Animal Behavior* 30:43–48.

Ressel, S.J. 1995. Ultrastructural properties of muscles used for call production in neotropical frogs. *Physiological Zoology* 69:952–973.

Richardson, C., J.-P. Lena, P. Joly, and T. Lengagne. 2008. Are leaders good mates? A study of call timing and male quality in a chorus situation. *Animal Behaviour* 76:1487–1495.

Ryan, M.J., R.B. Cocroft, and W. Wilczynski. 1990. The role of environmental selection in intraspecific divergence of mate recognition signals in the cricket frog, *Acris crepitans*. *Evolution* 44:1869–1872.

Ryan, M.J. and W. Wilczynski. 1988. Coevolution of sender and receiver: Effect on local mate preference in cricket frogs. *Science* 240:1786–1788.

Starnberger, I., D. Poth, P.S. Peram, S. Schulz, M. Vences, J. Knudsen, M.F. Barej, M.-O. Rödel, M. Walzl, and W. Hödl. 2013. Take time to smell the frogs: Vocal sac glands of reed frogs (Anura: Hyperoliidae) contain species-specific chemical cocktails. *Biological Journal of the Linnean Society* 110:828–838.

Taigen, T.L. and K.D. Wells. 1985. Energetics of vocalization by an anuran amphibian (*Hyla versicolor*). *Journal of Comparative Physiology B* 155:163–170.

Taigen, T.L., K.D. Wells, and R.L. Marsh 1985. The enzymatic basis of high metabolic rates in calling frogs. *Physiological Zoology* 58:719–726.

Taylor, R.C., B. Buchanan, and J. Doherty. 2007. Sexual selection in the squirrel treefrog *Hyla squirella*: The role of multimodal cue assessment in female choice. *Animal Behaviour* 74:1753–1763.

Taylor, R.C., B.A. Klein, J. Stein, and M.J. Ryan. 2008. Faux frogs: Multimodal signaling and the value of robotics in animal behavior. *Animal Behaviour* 76:1089–1097.

Toledo, L.F., I.A. Martins, D.P. Bruschi, M.A. Passos, C. Alexandre, and C.F.B. Haddad. 2015. The anuran calling repertoire in the light of social context. *Acta Ethologica.* 18:87–99.

Voituron, Y., L. Brepson, and C. Richardson. 2012. Energetics of calling in the male treefrog *Hyla arborea*: When being large means being sexy at low cost. *Behaviour* 149:775–793.

Wabnitz, P.A., J.H. Bowie, M.J. Tyler, J.C. Wallace, and B.P. Smith. 2000. Differences in the skin peptides of the male and female Australian tree frog *Litoria splendida:* The discovery of the aquatic male sex pheromone splendipherin, together with Phe8 caerulein and a new antibiotic peptide caerin 1.10. *European Journal of Biochemistry* 267:269–275.

Waldman, B. and P.J. Bishop. 2004. Chemical communication in an archaic anuran amphibian. *Behavioral Ecology* 15:88–93.

Weber, J.M. 2011. Metabolic fuels: Regulated fluxes to select mix. *Journal of Experimental Biology* 214:286–294.

Wells, K.D. 1977a. Territoriality and male mating success in the green frog (*Rana clamitans*). *Ecology* 58:750–762.

Wells, K.D. 1977b. The social behaviour of anuran amphibians. *Animal Behaviour* 25:666–693.

Wells, K.D. 1988. The effect of social interactions on anuran vocal behavior. In: Fritzsch B., M.J. Ryan, W. Wilczynski, T.E. Hetherington, and W. Walkowiak (eds.), *The Evolution of the Amphibian Auditory System*, pp. 433-454. New York: John Wiley and Sons.

Wells, K.D. 2007. *The Ecology and Behavior of Amphibians.* Chicago, Illinois: University of Chicago Press.

Wells, K.D. and C.R. Bevier. 1997. Contrasting patterns of energy substrate use in two species of frogs that breed in cold weather. *Herpetologica* 53:70–80.

Wells, K.D. and T.L. Taigen. 1986. The effect of social interactions on calling energetics in the gray treefrog (*Hyla versicolor*). *Behavioral Ecology and Sociobiology* 19:9–18.

Wells, K.D., T.L. Taigen, S.W. Rusch, and C.C. Robb. 1995. Seasonal and nightly variation in glycogen reserves of calling gray treefrogs (*Hyla versicolor*). *Herpetologica* 51:359–368.

Woodley, S. 2015. Chemosignals, hormones, and amphibian reproduction. *Hormones and Behavior.* 68:3–13.

Ziegler, L., M. Arim, and P. Narins. 2011. Linking amphibian call structure to the environment: The interplay between phenotypic flexibility and individual attributes. *Behavioral Ecology* 22:520–526.

Zimmitti, S.J. 1999. Individual variation in morphological, physiological, and biochemical features associated with calling in spring peepers (*Pseudacris crucifer*). *Physiological and Biochemical Zoology* 72:666–676.

chapter four

Digestive physiology in reptiles with special reference to pythons

**Sanne Enok, Lasse Stærdal Simonsen, Peter Funch,
Aksel Kruse, Jens Frederik Dahlerup, and Tobias Wang**

Contents

Introduction

All animals must eat as subsequent assimilation of the ingested food provides the energy and building blocks required to sustain all life functions. The functional performance of the digestive system therefore has implications for all physiological processes, and the ability to procure food, subdue prey, and retrieve its energy provides the basis for locomotion, growth, and reproduction (Wang 2001). Energy status therefore dictates the expression of most behaviors and while the ectothermic nature of reptiles implies that the energy devoted to basal life functions, that is, standard metabolic rate (SMR), is smaller than within endothermic birds and mammals, an effective digestive system is nevertheless needed to survive the long-lasting periods between suitable prey encounter. Such fasting periods, which may last for months in some species, are attended with rather impressive changes in the structure and function of the gastrointestinal (GI) organs (Pennisi 2005). This phenotypic flexibility seems to reduce the maintenance costs of the GI organs during fasting but is obviously only a viable strategy as long as the animals retain the capacity to swiftly upregulate digestive functions immediately upon prey ingestion; otherwise, the prey would deteriorate or even rot within the gut of the predator.

We base our review on the knowledge that has been accumulated on pythons, where particularly the Burmese and the ball pythons (*Python bivittatus* and *Python regius*, respectively) have been studied extensively in the past two decades. It is likely that the digestive and metabolic responses of the pythons, which are exquisitely well adapted to both very prolonged fasting and ingestion of very large meals, are considerably larger than in other reptiles. However, the basic mechanisms are probably shared among vertebrates.

Anatomy of the GI tract

The general structure of the reptilian GI tract resembles that of other vertebrates by being composed of an esophagus, stomach, pancreas, gall bladder, small intestine, appendix, large intestine, and cloaca. Nevertheless, the size of the total GI tract and its individual compartments can vary considerably depending on dietary specialization (Hume 1989; Alexander 1991; Horn and Messer 1992; Karasov and Hume 2010; Karasov et al. 2011). As a specialization to the many structural carbohydrates in the diet, herbivores tend to have a much longer digestive tract to increase gut retention time and maximize nutrient absorption. In contrast, carnivores readily absorb the easily digestible protein-rich meals with shorter guts and shorter retention times (Karasov et al. 2011).

Figure 4.1 shows the GI tract of a freshly euthanized fasting Burmese python, and also includes pictures taken *in vivo* by endoscopy as well as a photo of the open stomach (Figure 4.1e$_2$). The long and straight esophagus spans more than half of the GI tract and opens directly into the stomach without an apparent separation by a sphincter (Helmstetter et al. 2009a). However, *in vivo* endoscopy reveals several sphincter-like constrictions along the esophagus that may prevent regurgitation of stomach content. These constrictions were clearly dynamic in the live animal and probably reflect constriction of the circumferential muscle layers within the esophagus. Pythons ingest large prey, which resides within the esophagus before moving toward the stomach as the meal is progressively digested.

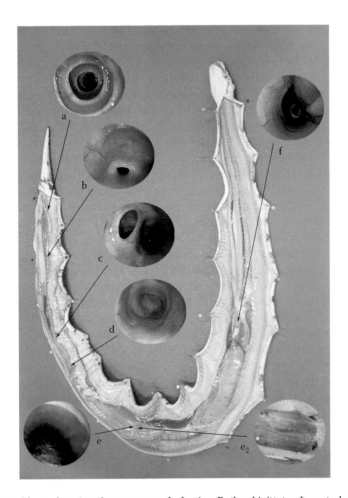

Figure 4.1 Photo showing the anatomy of a fasting *Python bivittatus.* Inserted pictures a–f show internal structures of the GI tract. All pictures are taken by an endoscope with the exception of insert e_2. Rectum (a), transition from the rectum to the sigmoid colon (b), sphincter from large to small intestine and opening to the appendix (c), ileocecal sphincter (d), transition from the esophagus to the stomach (e), external photo of the stomach lining (e_2), and heart and trachea seen from the esophagus (f).

The morphology of the gastric epithelia in pythons has been described in detail by Helmstetter et al. (2009a,b), and the division of a proximal fundic and distal pyloric region that terminates at the pyloric sphincter in the stomach is easily observed (Figure 4.1e + e_2). The wall of the fundus is thick, the mucosal epithelia are folded longitudinally, and gastric glands protrude into the gastric lumen. The upper neck region of the gastric glands contains mucus cells, while the lower pit region contains the oxyntopeptic cells, named after their shared secretion of both gastric acid and pepsinogen. These functions are separated into two specialized cells in mammals (Helmstetter et al. 2009a,b).

The mucosal epithelium of the pylorus is smooth with gastric glands appearing more as epithelial invaginations, as they lack the deep crypts of the gastric pits within the fundus. The pyloric glands are lined only with the mucus-secreting goblet cells (Helmstetter et al. 2009a,b). The pyloric sphincter opens into the small intestine, which is a long folded tube held in place by the mesentary. The intestinal epithelium is layered in a pseudostratified manner (Starck and Beese 2001; Lignot et al. 2005; Helmstetter et al. 2009a,b), which allows for the characteristic expansion of the epithelium in response to feeding. In contrast to mammals and birds, the small intestine does not contain crypts and has only a few goblet cells (Starck and Beese 2001). The pancreas and the gall bladder are located proximal to the pyloric sphincter, and the pancreatic and the bile ducts unite and empty into the intestine via a common duct (Moscona 1990). The junction between the small and large intestine is separated by a sphincter (Figure 4.1d) resembling the ileocecal valve in mammals. The large intestine has a folded luminal surface with the mucus-secreting goblet cells embedded in the intestinal epithelium (Helmstetter et al. 2009a). The appendix is located anterior to the sphincter and connected to the large intestine (Figure 4.1c). The passage from the sigmoid large intestine to the rectum is separated by a sphincter (Figure 4.1b), and there are several additional sphincters in the rectum before it terminates in the cloaca (Figure 4.1a).

Prey capture and ingestion

Foraging strategies vary considerably among reptiles but can be classified as either active foraging or ambush foraging (Greene 1997). The prey is normally swallowed intact immediately after capture with little or no mastication. In some species of snakes and a few lizards, venom is used to kill or immobilize the prey; they may also possess digestive properties, where proteases may initiate the breakdown from within the prey (Denburgh 1898; Alcock and Rogers 1902; McKinstry 1978; Thomas and Pough 1979; Fry et al. 2005; McCue 2005; Casewell et al. 2012), although more recent studies question the importance of this process (McCue 2007a; Chu et al. 2009).

Many groups of reptiles, including species of lizards, tortoises, and crocodilians, masticate their prey before ingestion, which aids digestion by breaking the food into smaller fragments, thus increasing the surface-to-volume area and making it more prone to enzymatic breakdown. The enzymatic degradation of food begins with the secretion of saliva from several glands in the oral cavity, which contains both amylases and lipases, initiating carbohydrate and fat digestion, respectively (Sanford 1992). In addition to digestive enzymes, the saliva also contains lysozyme, an antibacterial compound, and mucus, which eases swallowing (Sanford 1992). Little is known about the role of saliva in reptiles, but we recently observed very high concentrations of lysozymes in the saliva of pythons (Wang, unpublished), a finding that merits further clarification.

The ingesta proceeds to the stomach via the esophagus. To accommodate the extreme feeding behavior in many reptiles, the reptilian esophagus differs both anatomically and functionally from that of mammals (Ingelfinger 1958) with considerable specialization for large prey items (Uriona 2005; Cundall et al. 2014). In colubroid snakes, separation of the mandibles, rearrangement of the pharyngeal structures, and differences in smooth muscle layer associated with the anterior esophagus help in the swallowing of large prey (Cundall et al. 2014), while the esophagus of the American alligator has thicker musculature, composed only of smooth muscles and producing several fold larger peristaltic pressures (Uriona 2005).

Digestive process

The physiological responses within the GI tract can typically be divided into three distinct phases: cephalic phase, gastric phase, and intestinal phase. The cephalic phase begins before ingestion, as anticipation or the sight and/or smell of food initiate secretory processes and hence prepare the digestive organs prior to the actual presence of food within the GI tract (Johnson et al. 1991; Sanford 1992). The cephalic phase of gastric acid secretion is well known in mammals, but far less so in ectothermic vertebrates.

The gastric and intestinal phases are primarily mediated by ingested nutrients exerting a direct stimulant for secretion, increased motility, and activation of the absorptive processes (Johnson et al. 1991; Sanford 1992). Thus, the gastric phase of digestion is initiated when the ingesta reaches the stomach and stimulates the release of acid, proteases, and lipases both chemically and by mechanical distension of the stomach (Johnson et al. 1991; Sanford 1992). The proteolytic enzyme pepsin is released in an inactive form (pepsinogen) from gastric zymogen granules and activated in the acidic environment within the gastric lumen (Johnson et al. 1991; Sanford 1992; Cox and Secor 2008). The low luminal pH may also aid the breakdown of muscle and connective tissue of the prey and protect the GI tract from bacterial infections (Sanford 1992). Although the release of gastric acid would seem essential, pharmacological inhibition of acid secretion by the specific proton-pump inhibitor omeprazole does not affect the specific dynamic action (SDA)—the combined metabolic response to digestion—in pythons, pointing to the persistence of protein degradation in a neutral environment (Andrade et al. 2004). It would be interesting to characterize the pH sensitivity of the gastric proteases in snakes and other reptiles (Wang et al. 2002).

In pythons, gastric acid secretion ceases during the lengthy fasting period between meals, but the gastric pH decreases from 6.5 to 2 within 24 h after meal ingestion and continues to decline until the pH reaches approximately 1 (Secor and Diamond 1998; Secor 2003; Cox and Secor 2007). By mixing and grinding, the stomach also increases the surface area of the ingesta, enhancing the effects of the secreted enzymes and acid.

As the food particles are digested, the chyme moves to the proximal small intestine through the pyloric sphincter, thus commencing the intestinal phase of digestion. Gastric emptying is regulated by several hormones, but is also influenced by the size and liquidity of the ingesta (Jobling 1987; Low 2009). The low pH of the chyme may also inhibit gastric emptying (Low 2009). In reptiles, increased body temperature leads to increased rates of gastric digestion (Diefenbach 1975; Secor and Diamond 1995, 1997).

The continued breakdown of gastric chyme is ensured by the pancreatic and biliary secretions released in the proximal small intestine. The secretory duct differs in both number and amount, between and within species, dependent on the anatomy of the pancreas and the association with the spleen and gall bladder (Moscona 1990). In pythons, biliary and pancreatic secretions are released through a common duct (Moscona 1990).

Pancreatic juice contains bicarbonate and a combination of enzymes responsible for the breakdown of proteins, carbohydrates, and lipids (Johnson 1977; Sanford 1992). The bicarbonate neutralizes the acidic chyme, providing a suitable environment for the pancreatic enzymes and protecting the intestinal epithelium. In pythons, feeding triggers a 6-fold rise in activity of the pancreatic proteolytic enzyme trypsin and a 20-fold increase in pancreatic amylase activity within 4 days of feeding as food passes from the stomach to the intestine (Cox and Secor 2008).

The gall bladder secretes bile juice with bile salts, produced in the liver, that are important for assimilation, breakdown, and absorption of lipids. The conjugated bile alcohol or conjugated bile acids can be characterized depending on the length of the side chain and the terminal polar group (Hofmann et al. 2010). Although the different types of bile salts among classes can be used for phylogeny analyses, the physiological implication of these differences is not evident (Haslewood 1967; Moschetta et al. 2005; Hagey et al. 2010a,b; Hofmann et al. 2010). The bile ensures emulsification of the fat and the formation of micelles, which increase the surface area for lipases secreted from the pancreas. Although fat fails to elicit a metabolic response to digestion, a substantial decrease in gall bladder content upon feeding in pythons clearly indicates extensive lipid digestion (Hansen et al. 2013; Secor 1995, 2008; Secor et al. 2000b).

In infrequent feeders, enzyme capacities (product of enzyme activity and functional surface area) are down-regulated during fasting, which might minimize energy expenditure between meals (Cox and Secor 2008) but at the cost of being able to restore digestive enzyme capacities after feeding. This can be achieved by a combination of increased activity of existing enzymes, enzyme production, or functional surface area of the secretory organs (Cox and Secor 2008). Mammals primarily increase the specific activity of enzymes or increase enzyme production (Buddington and Diamond 1989; Ferraris et al. 1992), whereas pythons also increase the functional surface area, particularly of the small intestine (Starck and Beese 2001; Lignot et al. 2005).

As the small intestine is both secretory and absorbing, a structural and spatial gradient that mirrors the decrease in the luminal content of nutrients is present in the intestine (Cox and Secor 2008). The proximal small intestine has a greater mass, larger surface area, and larger uptake capacity than the distal small intestine (Secor and Diamond 1995; Lignot et al. 2005; Cox and Secor 2008). Enzyme activities decrease along the length of the intestine, corresponding with a decrease in functional surface area, the intestinal villi (Lignot et al. 2005; Cox and Secor 2008).

While lipids are passively absorbed in the distal small intestine, absorption of amino acids and carbohydrates occurs by both passive paracellular and carrier-mediated transcellular pathways (Karasov and Douglas 2013; Price et al. 2015). Paracellular transport is driven by the concentration gradient of water-soluble molecules that are sufficiently small to pass through the tight junctions between the enterocytes (Pappenheimer and Reiss 1987). This method of transport secures fast and energy-efficient transport of nutrients at the cost of selectivity, in contrast to carrier-mediated pathways, which work at a slower rate but with high selectivity (Pappenheimer 1993; Price et al. 2015). While mammals and large vertebrates rely mostly on transcellular transport, paracellular transport is important in small animals, providing a fast supply of nutrients to accommodate their high metabolic rates (Caviedes-Vidal et al. 2008; Karasov 2011; McWhorter et al. 2013). As the metabolic rate seems to be a factor in the ratio between paracellular and transcellular transport, lizards are believed to depend largely on transcellular transport, as selectivity of the ingested nutrients dominate over the velocity of nutrient uptake (McWhorter et al. 2013). However, it is hypothesized that diet may also have an effect on nutrient transport, as herbivores tend to rely less on paracellular transport compared to carnivores (McWhorter et al. 2013), possibly because herbivores are more exposed to toxins in their diet than carnivores, thus prioritizing selectivity (McWhorter et al. 2013). While paracellular transport is insaturable, carrier proteins must match intestinal performance to demand by increasing capacity in response to feeding (Karasov and Diamond 1983; Karasov et al. 1987; Secor and Diamond

2000; Secor 2005; Cox and Secor 2008). As with paracellular transport, the composition of carrier proteins depends on diet composition (Karasov and Diamond 1983, 1988; Karasov et al. 1987; Karasov 2011), which is also reflected in the notably larger uptake of amino acids compared to glucose in the strictly carnivorous pythons (Secor et al. 1994; Ott and Secor 2007; Cox and Secor 2008).

The chyme that proceeds to the large intestine is devoid of nutrients, but it remains imperative to reabsorb excess water and electrolytes (Johnson et al. 1991; Sanford 1992). The large intestine also functions as storage before defecation, which is aided by the secreted mucus (Johnson et al. 1991; Sanford 1992). In pythons, defecation, depending on meal size and temperature, takes place within 8–14 days after meal ingestion (Secor and Diamond 1995), notably longer than in mammals and other frequent eaters. The slow rate of digestion in reptiles reflects the lower metabolism and body temperatures (Diefenbach 1975). The interval between eating and defecating is particularly prolonged in terrestrial snakes, such as pythons and vipers, and is believed to function as a stabilizing ballast aiding in foraging (Lillywhite et al. 2002).

Hormones and regulatory peptides

The GI tract is the largest endocrine organ of the vertebrate body and releases vast amounts of peptides and hormones responsible for the initiation and regulation of the many digestive processes, ranging from the activity of exocrine glands to gut motility and the sensation of hunger or satiety (Rindi et al. 2004; Murphy and Bloom 2006; Simpson et al. 2008). The basic knowledge of these hormones comes primarily from mammals (Rindi et al. 2004; Murphy and Bloom 2006; Simpson et al. 2008), but some functions have been addressed in reptiles (Conlon et al. 1997a,b; Secor et al. 2001). Most hormonal actions appear to be highly conserved among vertebrates, but it remains problematic that the infusion of mammalian antagonists and agonists may yield erroneous responses because of structural differences in the amino acid sequences in other animals. Fortunately, the recent sequences of reptilian genomes will greatly alleviate these problems.

The hormonal versus the luminal components of the intestinal phase in pythons was elegantly studied by Secor and colleagues (Secor et al. 2000b, 2002). In these snakes, a part of the small intestine was surgically separated from the rest of the gut several weeks before voluntary feeding. This isolated gut section, which continued to receive normal perfusion and hence the normal suite of blood-borne hormones and regulatory signal molecules, as well as its normal nervous innervation, showed a delayed, but nevertheless notable, upregulation of nutrient transport capacity in response to feeding, while the expansion of the mucosa and the growth of microvilli was virtually absent. These interesting findings point to the rise in nutrient transport capacity being surprisingly uncorrelated to the morphological response of the intestinal lining.

In mammals, the gastric mucosa releases ghrelin, gastrin, histamine, and somatostatin (Rindi et al. 2004). Ghrelin increases appetite (Murphy and Bloom 2006; Simpson et al. 2008), whereas gastrin, secreted from G cells in the lower gastric mucosa, provides a direct stimulation of gastric-acid-secreting cells and an indirect stimulation through the release of histamine from enterochromaffin-like (ECL) cells (Rindi et al. 2004; Murphy and Bloom 2006). Gastrin could not be identified in the python gastric mucosa using antibodies, probably not because acid secretion is regulated differently than in mammals, but most likely because the structure of python gastrin differs from mammals. Somatostatin is released from gastric D cells, as well as the intestine and pancreas, and is thought to have an inhibitory effect on gastric and pancreatic secretions, gut motility, gall bladder contractions, and absorption of amino acids, although it is not well described (Secor et al. 2001). With the exception of histamine, these gastric hormones are also found throughout the intestinal mucosa (Rindi et al. 2004; Murphy and Bloom 2006).

The intestinal mucosa releases a plethora of hormones from distinct endocrine cells: serotonin, neurotensin, motilin, cholecystokinin (CCK), glucose-dependent insulinotropic polypeptide (GIP), glucagon-like-peptide-1 and 2 (GLP-1 and GLP-2), secretin, and peptide YY (PYY) (Rindi et al. 2004). In humans, ECL cells are located throughout the intestine, releasing serotonin (Rindi et al. 2004). The function of serotonin is not well known but seems to affect appetite (Murphy and Bloom 2006). Motilin affects gastric motility, while secretin stimulates the pancreatic secretion of acid-neutralizing bicarbonate (Rindi et al. 2004; Murphy and Bloom 2006). CCK is released from the upper part of the small intestine in response to feeding, especially by the presence of amino acids and long-chain fatty acids, acting to reduce appetite when binding to the vagal CCK1 receptor (Rindi et al. 2004; Murphy and Bloom 2006; Simpson et al. 2008). CCK also stimulates pancreatic growth and secretion and induces gall bladder contractions (Murphy and Bloom 2006). The postprandial concentrations of CCK increase 3–4-fold in humans (Secor et al. 2001), whereas pythons, because of their ability to ingest very large meals and their long fasting intervals, experience a 25-fold rise (Secor et al. 2001). GLP-2 stimulates GI motility, absorption, and growth (Murphy and Bloom 2006). PYY is secreted primarily in the distal small intestine and the large intestine from L-cells in response to feeding, acting to reduce appetite and gut motility, often referred to as the ileal break (Lin et al. 2004; Rindi et al. 2004; Murphy and Bloom 2006; Simpson et al. 2008).

The endocrine pancreas releases hormones important for glucose homeostasis (Rindi et al. 2004; Murphy and Bloom 2006; Simpson et al.

2008). The pancreatic β-cells release amylin and insulin. Amylin is released in response to feeding, reducing appetite, and maintaining glucose homeostasis (Murphy and Bloom 2006; Simpson et al. 2008). Insulin is an anabolic hormone released primarily in response to glucose, acting to increase glucose uptake in lean and adipose tissue and inhibiting gluconeogenesis (Murphy and Bloom 2006; Simpson et al. 2008). In contrast, during fasting, glucagon is released from pancreatic α-cells and maintains blood glucose via gluconeogenesis and breakdown of glycogen stores (Marieb and Hoehn 2007). However, it appears that glucagon levels increase during digestion in pythons (Secor et al. 2001), which may indicate an entirely different function of this hormone and may serve to elevate plasma glucose to fuel the digestive process before ingested energy becomes available (Secor et al. 2000b). Nevertheless, it is likely that the effects are determined by the glucagon:insulin ratio, as in other vertebrates, and this mechanism clearly warrants further investigation. Pancreatic polypeptide is also released from the pancreas and reduces appetite and gastric motility (Murphy and Bloom 2006; Simpson et al. 2008).

Insulin secretion is regulated by two other gut hormones, the incretins, GIP and GLP-1, that augment the glucose-stimulated insulin secretion in mammals (Rindi et al. 2004; Murphy and Bloom 2006; Simpson et al. 2008; Holst et al. 2009). Pythons produce GIP in the pancreas (Secor et al. 2001), as opposed to the small intestine in humans, which might indicate a different function of the hormone. GIP in pythons thus seems to have a different incretin effect, if any at all, possibly relating to the fact that pythons are strictly carnivorous. GLP-1 also reduces appetite and gut motility (Murphy and Bloom 2006; Simpson et al. 2008).

Digestive effects on visceral organs

The availability of food often varies with seasonal or geographical changes. Animals typically reproduce when food is abundant, whereas they reduce activity or migrate when food availability is scarce. When migration is not an option, animals depend entirely on internal energy stores until food again becomes available, and in this case, a reduction in energy expenditure will serve to match demand to the absence of energy supply (Wang et al. 2006).

Reduction of physical activity, reproduction, and growth are all common ways to reduce overall energy expenditure. Lowering of body temperature is a common strategy to reduce standard metabolism and hibernation; estivation or daily torpor commonly depresses energy expenditure by 80% in vertebrates; and some bats even decrease metabolism by more than 95% (Humphries et al. 2003; Geiser 2004; Bicego et al. 2007; Navas and Carvalho 2010; Geiser and Stawski 2011). While some reptiles and amphibians also enter dormancy or move to colder niches (Abe 1995; Guppy and Withers 1999; Milsom et al. 2008; Toledo et al. 2008; Campen and Starck 2012), this is not a common tactic (Piersma and Lindström 1997; Wang et al. 2006). Reduction in standard metabolism may therefore be an important strategy (Campen and Starck 2012), although subject to conflicting findings. One hundred sixty-eight days of starvation in rattlesnakes resulted in a 24% loss in body mass along with a 72% decrease in metabolism; however, recent data on rattlesnakes that lost 30% of body mass during one entire year of fasting did not show a reduced resting metabolic rate (McCue 2007b; Leite et al. 2014). Also, 112 days of starvation in pythons resulted in an 18% loss in body mass along with a 41% decrease in metabolism (McCue 2007b), but in a previous study, no changes in standard metabolism were found after 60 days fasting (Overgaard et al. 2002).

The maintenance of the GI system is considered energetically expensive and has been estimated to account for up to 20%–30% of SMR in mammals (Stevens and Hume 2004), and an even higher contribution has been proposed in reptiles (Secor and Diamond 2000). While most organs are continuously active, the GI tract remains quiescent between meals, and a down-regulation of these organs could therefore decrease metabolism while maintaining other bodily functions. This mechanism would be particularly advantageous in intermittent feeders, such as many reptiles, that experience prolonged fasting, and several-fold reductions in intestinal mass is common among fasting crocodilians and snakes (Secor et al. 1994; Secor and Diamond 1995, 1998; Starck et al. 2007).

The pronounced morphological restoration of the GI organs upon feeding was initially believed to be energetically expensive (Secor and Diamond 1998; Secor et al. 2000a), and the high energetic cost of digestion (SDA) of infrequently feeding snakes was attributed to cell proliferation during restoration of the gut (Secor et al. 2000a,b). However, it is now generally agreed that the morphological changes primarily involve expansion of the individual enterocytes that appear to occur at modest energy expenditure (Overgaard et al. 2002). It is now established that the intestine expands primarily by hypertrophy, that is, increased size of the individual enterocytes, possibly by incorporation of lipid droplets after feeding, such that the intestinal surface is unfolded (Starck and Beese 2001; Starck 2002; Starck et al. 2007). In pythons, this plasticity is at least partly due to pseudostratified epithelia found throughout the gut (Helmstetter et al. 2009b), but the rise in surface area is also attributed to a 2-fold lengthening of the microvilli (Figure 4.2). In addition to the intestine, other organs such as the stomach, liver, and pancreas also show reversible changes in mass (Figure 4.3) (Enok et al. 2013).

While reducing the size of the energy-consuming intestine appears to be advantageous during long periods of fasting, it is obviously imperative that the phenotypic response is reversible, so the animal can initiate

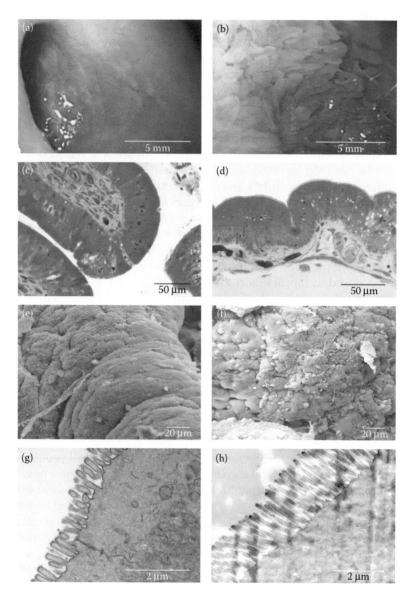

Figure **4.2** Postprandial growth and modifications of the small intestine in pythons fed a meal of 25% of BW. Endoscopic pictures from the small intestine in a fasted (a) and fed (b) python. Light microscopy of thin sections of toluidine blue stain plastic embedded small intestinal villi from a fasted (c) and fed (d) python. Scanning electron micrographs of the surface of small intestinal enterocytes from a fasting (e) and fed (f) python. Transmission electron micrographs of the microvilli in the small intestine in fasting (g) and fed (h) pythons. (From Funch, P, T Wang, L S Simonsen, S Enok, J F Dahlerup, and A Kruse. Unpublished).

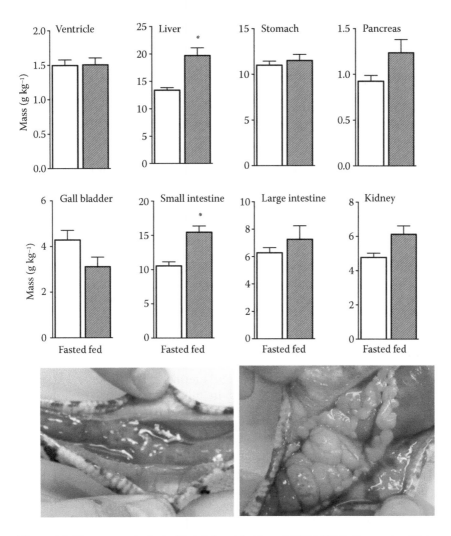

Figure 4.3 Organ mass in fasted (white) and fed (gray) (18% BW) *Python regius*. (Data adapted from Enok, S, L S Simonsen, and T Wang. 2013. *Comparative Biochemistry and Physiology Part A: Molecular & Integrative Physiology* 165 (1): 46–53. doi: 10.1016/j. cbpa.2013.01.022.) Inserted pictures illustrate the postprandial small intestinal growth in one *Python bivittatus*. Values are mean ± standard error of mean (SEM). * Denotes significant difference from fasting values.

digestive functions immediately after feeding. In snakes, intestinal mass increases already within 12 h after feeding, doubles at 24 h, and maximizes some 2–3 days after ingestion (Secor and Diamond 1995; Starck et al. 2007). Intestinal mass remains elevated until defecation and then regresses progressively (Secor and Diamond 1998). The signal triggering growth of the intestine and other GI organs could either be a direct effect of feeding, hormonal signaling, or a secretory response (Wang et al. 2006). Surgically isolating a small part of the small intestine in juvenile pythons showed that the trigger for postprandial organ growth in pythons is luminal (Secor et al. 2000b). The isolated piece of intestine was placed next to the reanastomosed intestinal segments with intact mesenteric nerves and blood supply. The reanastomosed intestine doubled in mass within 24 h after feeding, while the isolated intestinal segment, receiving no luminal signals, showed no morphological changes. The snakes were fed meals by injections of fat, glucose, protein, bile, or whole rodent to the intestine, but only protein and whole rat induced a response, indicating that changes in python intestinal mass are due to the direct presence of protein in the gut (Secor et al. 2000b).

Metabolic response to digestion

Metabolism increases several fold during digestion, and this postprandial rise in metabolism is caused by preabsorptive, absorptive, and postabsorptive processes, namely, all processes from prey handling to postabsorptive protein synthesis.

Specific dynamic action

This metabolic response to digestion is termed SDA (Rubner 1902; Lusk 1922) and has been characterized in all major groups of vertebrates (Andrade et al. 2005; McCue 2006; Secor 2009). The SDA response differs considerably between species, with peak responses ranging from 25% in humans to more than 600% in snakes such as pythons (Secor 2009). The large responses in pythons are due to the large meals, but their factorial rise in metabolism may also be larger than frequent feeders that do not downregulate the gut during fasting (Secor 1997; Secor and Diamond 1998).

The SDA response is characterized by an immediate increase in oxygen consumption after a meal followed by a slower decline, to prefeeding levels, a response that can last up to 2 weeks. The cost of the SDA response can be expressed as the total amount of oxygen consumed, but often the SDA coefficient (SDA%) is used. SDA% expresses the cost of digestion as a percentage of the energy ingested, and the cost of different meal sizes and compositions can then readily be compared. The SDA response is often divided into three overlapping phases: preabsorptive, absorptive, and postabsorptive, or the cephalic, gastric, and intestinal phase. The preabsorptive phase includes many processes such as prey handling, gut peristalsis, acid and enzyme secretion, and intestinal remodeling. The absorptive processes are mainly characterized by nutrient transport and intestinal absorption, while postabsorptive processes include glycogen and urea production, renal excretions, protein synthesis, and growth (McCue 2006). The specific energetic contribution of each of these processes remains debated and largely unknown, mostly because the different phases and processes overlap temporarily and depend on each other, making a single process difficult to distinguish from one another. However, several studies investigating the digestive components have been made.

The cephalic phase, part of the preabsorptive phase, has been investigated in snakes several times. The cost of constriction, inspection, and swallowing of the prey in *Boa constrictor* was estimated by measuring oxygen consumption during this preabsorptive phase of digestion. These processes represent less than half a percent of the total energy in the meal (Cruz-Neto et al. 1999; Canjani et al. 2002), a minor part of the total cost of an SDA (for pythons this accounts for ~30% of the ingested energy). This cost also includes the repayment of the oxygen debt in the inspection phase, due to possible anaerobic work during constriction of the prey (Cruz-Neto et al. 1999; Canjani et al. 2002). In pythons, the cost of the cephalic phase was estimated by letting the snake constrict and inspect the prey before removing it from the snake, showing only a minor contribution to the SDA response (Secor et al. 2002). Although the cephalic phase is energetically inexpensive, other preabsorptive gastric processes may contribute (Secor 2003). Gastric digestion primarily involves the mechanical and chemical degradation of food and it has been hypothesized that larger and more intact meals would be more costly to digest than liquid and masticated meals (Secor 2009). Thus, homogenization of the prey (a rat) reduced the SDA response by 26%, and infusion of a homogenized rat directly into the intestine caused a 67% reduction. On this basis, gastric work was estimated to account for up to 55% of the total SDA response (Secor 2003). However, pharmacological inhibition of the gastric acid secretion with omeprazole had no effect on the SDA response, indicating that gastric acid secretion is an inexpensive process (Andrade et al. 2004). In addition, Henriksen et al. (2015) and Nørgaard et al. (2016) took the opposite approach stimulating gastric acid secretion in pythons, by the addition of bone meal to a protein rich meal – increasing buffer capacity of the meal. The SDA response was unchanged, indicating gastric acid secretion is of modest metabolic cost, supporting Andrade et al. (2014). Although gastric acid secretion is an important component of gastric digestion, it does not entirely exclude the possibility that the major cost of gastric processes could stem from the mechanical part of digestion. To investigate the cost of the combined gastric processes, Enok et al. (2013) performed complete pyloric separation (Enok et al. 2013) to prevent the chyme from reaching the intestine, and reported a pronounced lowering of the initial SDA response, suggesting that the gastric processes are indeed a minor part of the SDA.

If the cephalic phase, acid secretion, and combined gastric processes are energetically inexpensive, the pronounced rise in oxygen consumption must stem from either the absorptive or postabsorptive phase. Although the absorptive phase accounts for many processes (gut peristalsis, nutrient transport, nutrient absorption, enzyme action, and production), the cost is thought to be minor (Secor 2009). Some investigators have, however, hypothesized that the intestinal phase is of more importance in pythons, due to a greatly increased blood flow to the intestine during digestion. This indicates increased energy expenditure by the intestinal tissue (Secor 2005), but most recent studies agree that the intestinal remodeling during digestion is energetically cheap (Starck and Beese 2001; Overgaard et al. 2002). The causes of the SDA response remain subject to conflicting views. While some advocate the preabsorptive phase as a major component (Secor et al. 2012), others believe the SDA response in snakes is primarily due to the postabsorptive protein synthesis (Wang et al. 2006; Enok et al. 2013). Supportive of the latter hypothesis, inhibition of protein synthesis reduced the SDA response by 70% in snakes (McCue et al. 2005). Intravenous infusion of amino acids also increases the metabolism to the same magnitude as if consumed orally in dogs, catfish, and plaice (Brown and Cameron 1991a,b). The SDA response is affected by numerous parameters, and both magnitude and duration of the response show great plasticity. Features of the meal, such as composition, size, and type, affect SDA, as do other variables, such as feeding frequency, temperature, sex, age, body type, and composition.

Effects of temperature

Temperature affects most physiological processes, and ectotherms are entirely dependent on the ambient temperature for their physiological functions, such as locomotion and digestion (Brett 1971; Fry 1971; Dawson 1975; Harlow et al. 1976; Huey and Stevenson 1979; Huey and Hertz 1984). Snakes prefer to digest at higher temperatures and will even regurgitate if temperatures fall below a lower critical temperature (Wang et al. 2002; Tsai et al. 2008). The shape of the SDA response is markedly affected by temperature. In general, SMR and the maximal oxygen uptake ($\dot{V}O_2$ peak) increase with increasing temperatures (Wang et al. 2002; Toledo et al. 2003, 2008; Bessler et al. 2010), while the duration and the time to reach $\dot{V}O_2$ peak decrease (Toledo et al. 2003, 2008; Bessler et al. 2010). An example in Figure 4.4a shows the oxygen uptake of digesting pythons at 20°C, 25°C, 30°C, or 35°C (Wang et al. 2002). Both SMR and $\dot{V}O_2$ peaks increase with temperature, yet the factorial increase in postprandial metabolism does not necessarily change. The increase in both SMR and $\dot{V}O_2$ peaks seems similar in response to increasing temperatures in pythons, thus yielding the same factorial increase in postprandial metabolism (Wang et al. 2002). In contrast, other snakes have optimum temperatures, where digestive processes are ideal. The Chinese green tree viper

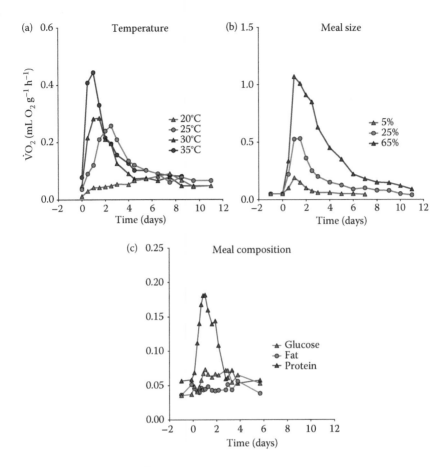

Figure 4.4 (a) The mean V̇O₂ of *P. bivittatus* at different temperatures (20°C, 25°C, 30°C, and 35°C) from time of feeding (20% BW at 0 h) to 11 days after feeding. (Adapted from Wang, T et al. 2002. *Comparative Biochemistry and Physiology Part A—Molecular & Integrative Physiology*. 133: 519–27.) (b) The mean V̇O₂ of *P. bivittatus* fed meals of different sizes (5%, 25%, and 65% of BW). Data are shown from 1 day prior to feeding until 11 days after feeding. (Data adapted from Secor, S M and J Diamond. 1997. *The American Journal of Physiology* 272 (3 Pt 2) R902–12.) (c) Mean V̇O₂ of *P. regius* fed meals of different compositions (glucose, fat, or protein) equivalent to approximately 8% of BW. Data are shown from 1 day prior to feeding and for the following 7 days. (Adapted from Henriksen PS et al. 2015. *Comparative Biochemistry and Physiology Part A: Molecular & Integrative Physiology*. 183:36–44.)

(*Trimeresurus stejnegeri stejnegeri*) has an optimum temperature of 25°C where the factorial increase in metabolism is larger than at both higher and lower temperatures (Tsai et al. 2008), whereas the garter snake shows a significant factorial increase when the temperature increases from 15°C to 20°C (Bessler et al. 2010). Increasing temperatures are therefore not always equal to a rise in the factorial metabolic increase due to digestion. The total cost of digestion, expressed as both the SDA and the SDA%, is also subject to conflicting findings. SDA and the SDA% have been found not to differ with increasing temperature in the python and the Chinese green tree viper (Tsai et al. 2008), whereas both parameters decrease in boas (Toledo et al. 2003). Garter snakes differ by increasing SDA and SDA% in response to increasing temperature (Bessler et al. 2010).

Effects of meal size and composition

Snakes can ingest meals exceeding their own mass, and although normal meals are typically in the range of 10%–30% of their body mass, these large meals impose a challenge on the GI organs to digest the meal as fast as possible. Retention of bulky foods in the stomach increases the possibility of predation and may thus reduce overall fitness (Garland and Arnold 1983); so, fast digestion is ecologically important. Meal size has indeed been shown to increase the metabolic response during digestion (Andrade et al. 1997; Secor and Diamond 1997; Tsai et al. 2008; Bessler et al. 2010), demonstrated by a progressive rise in peak postprandial oxygen consumption with increasing meal size as illustrated in Figure 4.4b where pythons were fed meals of either 5%, 25%, or 65% of body weight (BW) (data from Secor and Diamond [1997]). Overall, the response to a doubling in food intake has been estimated to contribute to a more than 50% increase in peak oxygen consumption and a 30% increase in SDA duration resulting in a more than doubled SDA (Secor 2009). Similar results have been found in Chinese green tree viper (Tsai et al. 2008). The metabolic response to increased meal size reaches a plateau in species such as fishes, hypothesized to be due to a maximum oxidative capacity of the gut tissue (Jobling and Davies 1980). Snakes, however, do not seem to experience this plateau. Oxygen uptake continuously increases after meals as large as 100% of BW (Andrade et al. 1997; Secor and Diamond 1997; McCue and Lillywhite 2002; Toledo et al. 2003). The size effect of the meals seems to be primarily due to mass and composition, since the volume of the meal does not correlate with SDA (McCue et al. 2005). Likewise, balloon distention of the stomach does not seem to elicit a digestive response, since postprandial intestinal uptake capacity and intestinal morphology were similar to that found in fasting snakes (Secor et al. 2002).

Meal composition also affects the SDA response. In snakes, ingestion of fat does not increase oxygen consumption, while the SDA response increases with increasing amounts of protein (Figure 4.4c) (Coulson and Hernandez 1979; McCue et al. 2005). Larger carbohydrates, such as cellulose and starch, are not readily digested in pythons (McCue et al. 2005). Only small sugars such as glucose and sucrose are digested and show a somewhat similar metabolic response compared to protein (McCue et al. 2005).

The type of meal is also important in determining the response. Simple proteins do not elicit similar responses as do more complex proteins. A complete mixture of amino acids also elicits a larger response than does a less complete mixture. However, the most pronounced increase in post-prandial metabolism is found when snakes are fed a complete meal, such as, a rodent, although an increasing relative amount of protein results in an increased SDA response (Secor et al. 2002). This trend is consistent with the theory of protein synthesis as the major contributor to the SDA response.

Fueling the metabolic response

The SDA response must be fueled, either from the energy ingested or from the energy stored in the body, prior to the assimilation of nutrients. The rapid increase in oxygen consumption indicates that internal stores must be the initial fuel (Secor and Diamond 1998; Secor et al. 2001; Secor 2003), which prompted the term "pay-before-pumping" (Secor and Diamond 1995). The hypothesis that the snakes incurred initial costs during digestion was supported by the fact that plasma triglycerides increased 160-fold within 24 h of digestion and that plasma glucagon levels increased after feeding, correlating with the peak in oxygen consumption (Secor and Diamond 1998).

To study how much of the reliance was on endogenous energy stores (i.e., own resources) versus the exogenous energy sources derived from the prey, Waas et al. (2010) fed ^{13}C-labeled prey to snakes. Within 4 h after ingestion, the ^{13}C appeared in the exhaled air and rose steadily and accounted for 75% of total respiration by 24 h into the digestive process (see Figure 4.5; Waas et al. 2010). It was estimated that the prey contributes approximately 60% to the entire SDA response. The relatively high reliance on the prey to fuel the SDA response is consistent with the rate of digestion estimated in previous experiments. Thus, after 24 h of digestion, 15%–27% of the stomach content was passed to the intestine for a meal of 25% of BW (Secor 1995; Secor and Diamond 1995, 1997) and could therefore readily be utilized for fueling digestion. The increased plasma concentration of triglycerides, from the digested fatty acids might instead be part of the hypertrophic response to digestion (Starck and Beese 2001).

Cardiovascular response to digestion

The substantial rise in metabolism during digestion is supported by an equally large cardiovascular response (Starck and Wimmer 2005; Secor 2008; Wang and Skovgaard 2008). Cardiac output, and particularly flow in the mesenteric artery, increase several fold by concomitant increases in both heart rate, stroke volume, and decrease in systemic resistance

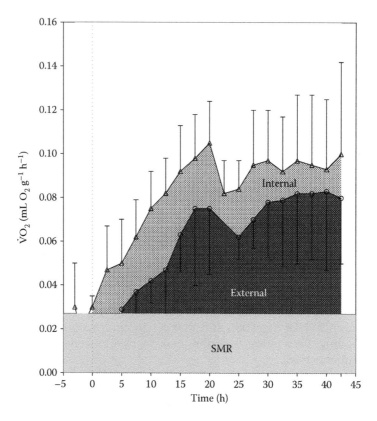

Figure 4.5 Postprandial oxygen consumption in *P. regius* fed a ^{13}C-labeled rodent meal corresponding to 8% BW. Triangular symbols indicate the total $\dot{V}O_2$ during 43 h after feeding. The light gray area is the portion of the SDA fueled by endogenous energy sources. The dark gray area represents the portion of the SDA fueled by oxidizing exogenous energy sources (^{13}C-labeled rodents). Mean SMR is indicated by the horizontal gray area. Data are shown from time of feeding (0 h—dotted line) and for the following 43 h and is represented as mean ± SD. (Adapted from Waas, S, R A Werner, and J M Starck. 2010. *Journal of Experimental Biology* 213 (8): 1266–71. doi: 10.1242/jeb.033662.)

(Figure 4.6b–d) (Secor et al. 2000a,b; Starck and Wimmer 2005), while blood pressure is maintained (Wang et al. 2001b; Skovgaard et al. 2009; Enok et al. 2012). This effect of digestion is represented in Figure 4.6a showing a trace of heart rate and blood pressure measurements from a boa before and after feeding.

The intestinal hyperemia has been studied in fishes, reptiles, and mammals, and they all show large increases in the intestinal blood flow at magnitudes of 50%–150% (Rees et al. 1982; Perko et al. 1998; Axelsson et al. 2000; Madsen et al. 2006; Altimiras et al. 2008; Eliason et al. 2008;

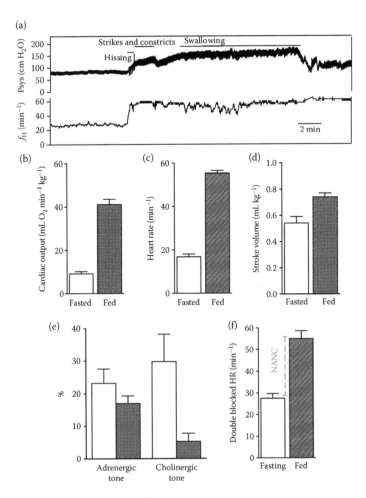

Figure 4.6 Top panel shows blood pressure and heart rate trace from a boa before and during presentation of a meal. The trace illustrates increases in both blood pressure and heart rate during the phases of meal consumption (Data adapted from Wang, T, E Taylor, and D Andrade. 2001b. *The Journal of Experimental Biology* 204: 3553–60). (b–d) Cardiac output, heart rate, and stroke volume from *P. bivittatus* fed a rodent meal equivalent to 25% of BW (30°C). Data are presented from time of feeding (0 h) until 15 days into the postprandial period. All data are mean ± SEM. (Data adapted from Secor, S M and White S E. 2010. *The Journal of Experimental Biology* 213: 78–88. (e) Adrenergic and cholinergic tone in postprandial *P. bivittatus* (30°C), where white is fasted and gray is fed. (f) Double blocked (atropine and propanalol) heart rate in *P. regius* fed a meal of 25% of BW (30°C). The difference in double blocked heart rate of fasted and fed snakes illustrate the postprandial NANC factor. (Data adapted from Skovgaard, N et al. 2009. *American Journal of Physiology: Regulatory, Integrative and Comparative Physiology* 296 (3): R774–85. doi: 10.1152/ajpregu.90466.2008.) All data are mean ± SEM.

Seth et al. 2009). However, pythons, after a meal of 25% BW, increase blood flow up to 11 folds in the superior mesenteric artery, supplying intestinal blood flow (Secor et al. 2010). The much more pronounced increase in blood flow in pythons is most likely due to the much larger increase in postprandial oxygen consumption compared to other vertebrates (Wang and Skovgaard 2008). The increased intestinal blood flow can be achieved a redistribution of blood flow from other organs, which occurs when vascular resistance is altered, thereby moving blood from areas with high resistance to areas with low resistance. Redistribution of blood flow is controlled by several factors acting on vascular tone such as neural mediators, circulating humoral factors, circulating paracrine and autocrine mediators, and metabolic vasodilators (Starck and Wang 2005). During digestion, several local metabolites such as substance P, neuropeptide Y, and neurotensin are released (Secor et al. 2001). These substances have been shown to decrease intestinal vascular resistance in pythons and thus mediate an increased intestinal blood flow during digestion (Wang et al. 2000; Skovgaard et al. 2005, 2007). The decreased vascular resistance in pythons is accompanied by a more than 4-fold increase in cardiac output (Figure 4.6b) (Hicks et al. 2000; Secor et al. 2000a).

Cardiac output is the product of heart rate and stroke volume, both of which can be modulated by the autonomic nervous system. The stroke volume of reptiles is relatively unaffected by body temperature or exercise, and large changes in cardiac output are primarily due to changes in heart rate (Hicks et al. 2000; Starck and Wang 2005). The sympathetic limb increases heart rate and contractility, whereas the parasympathetic innervation slows the heart (Taylor et al. 1999). In pythons, stroke volume may rise by 50% during digestion which, in combination with the attending tachycardia, causes a 4-fold rise in cardiac output during digestion (Secor et al. 2000a; Secor 2008) (Figure 4.6b–d). The postprandial tachycardia in pythons is accompanied by a decrease in parasympathetic tone, while the sympathetic tone is either slightly decreased or unaltered (Figure 4.6e) (Cox et al. 1995; Wang et al. 2001b). In addition to the autonomic nervous system, the postprandial heart rate is regulated by non-adrenergic-non-cholinergic (NANC) factors, revealed after a double autonomic blockade of the heart (Figure 4.6f) (Wang et al. 2001b; Skovgaard et al. 2009; Enok et al. 2012).

In addition to the postprandial growth of the GI organs (Secor 2008; Jensen et al. 2011), there are several reports of reversible increases of heart mass in pythons (Secor and Diamond 1995; Secor and Diamond 1997; Andersen et al. 2005; Riquelme et al. 2011), which could explain the rise in stroke volume during the postprandial period. A number of recent studies, however, failed to confirm this growth (Jensen et al. 2011; Enok et al. 2013). To understand the discrepancy between these findings, Slay et al. (2014) investigated the possibility that a lowering of venous oxygen levels provide a trophic hypoxic signal to the heart tissue and showed that there was cardiac remodeling during digestion in anemic snakes (Slay et al. 2014). Thus, it is quite possible that low venous oxygen levels exert a trophic signal that stimulates cardiac growth.

Blood oxygen affinity and the alkaline tide

Blood oxygen affinity and carrying capacity also affect oxygen delivery to the digestive tissues. Blood oxygen-carrying capacity is mostly determined by hemoglobin concentration, but most studies show rather small increases in hematocrit during digestion. The issue, however, is somewhat complicated because such data are obtained from experimental animals in which repeated blood sampling may have reduced hematocrit (Wang et al. 1995; Overgaard et al. 1999; Busk et al. 2000a,b; Wang et al. 2001a). Two studies on snakes (*Python* and *Crotalus*) show that blood oxygen affinity increases slightly during digestion (Overgaard and Wang 2002; Bovo et al. 2014), but it remains to be studied whether these modest effects serve to improve oxygen transport.

The secretion of gastric acid where there is a net transfer of HCl to the stomach lumen causes a rise in HCO_3^- as chloride is supplied by blood supplying the oxyntopeptic cell. This has been termed the "alkaline tide," but in all reptiles studied, there is actually virtually no alkalinization of the body fluids because arterial PCO_2 increases, such that the pH remains constant (Overgaard et al. 1999; Busk et al. 2000a,b; Arvedsen et al. 2005). The digestive state is therefore characterized by a respiratory compensation of the metabolic alkalosis caused by acid secretion. The respiratory compensation arises because ventilation does not increase as much as metabolic CO_2 production during digestion.

Conclusions

The function and morphology of the digestive organs of snakes exhibit extraordinarily pronounced phenotypic flexibility in response to variations in food availability. The rapid upregulation of the GI functions appears to occur at relatively small energy expenditure, but there must obviously be energetic costs associated with replenishing the digestive organs upon digestion, such that enzymes, enterocytes, and ions as well as nutrient transporters are ready for the coming meal. Much of these processes remain to be understood in more detail and the recent molecular studies addressing gene expression profiles during fasted and fed conditions may soon begin to allow more detailed analysis of these processes at the cellular level.

Because of the large changes in metabolism during digestion, the postprandial period represents an integrated response involving virtually all organ systems that require exquisite regulation at several levels of biological organization. Many of the interactions between organ systems have been studied in laboratory experiments, but given the rapid advance in smaller and more sophisticated data loggers with measurements of numerous physiological parameters, it is likely that we will soon be able to record to what extent the various physiological adaptations translate into improved performance in animals behaving freely under natural conditions.

References

Abe, A S. 1995. Estivation in South American amphibians and reptiles. *Brazilian Journal of Medical and Biological Research = Revista Brasileira De Pesquisas Médicas E Biológicas/Sociedade Brasileira De Biofísica ... [Et Al.]* 28 (11–12): 1241–47.

Alcock, A and L Rogers. 1902. On the toxic properties of the saliva of certain "non-poisonous" colubrines. *Proceedings of the Royal Society of London* 70: 446–54.

Alexander, R M. 1991. Optimization of gut structure and diet for higher vertebrate herbivores. *Philosophical Transactions of the Royal Society B: Biological Sciences* 333 (1267): 249–45. doi: 10.1098/rstb.1991.0074.

Altimiras, J, G Claireaux, E Sandblom, A P Farrell, D J McKenzie, and M Axelsson. 2008. Gastrointestinal blood flow and postprandial metabolism in swimming sea bass *Dicentrarchus labrax. Physiological and Biochemical Zoology: PBZ* 81 (5): 663–72. doi: 10.1086/588488.

Andersen, J B, B C Rourke, V J Caiozzo, A F Bennett, and J W Hicks. 2005. Physiology: Postprandial cardiac hypertrophy in pythons. *Nature* 434 (7029): 37–38. doi: 10.1038/434037a.

Andrade, D V, A P Cruz-Neto, and A S Abe. 1997. Meal size and specific dynamic action in the rattlesnake *Crotalus durissus* (Serpentes: Viperidae). *Herpetologica* 53 (4): 485–93.

Andrade, D V, A P Cruz-Neto, A S Abe, and T Wang. 2005. Specific dynamic action in ectothermic vertebrates: A review of the determinants of postprandial metabolic response in fishes, amphibians, and reptiles. In *Physiological and Ecological Adaptations to Feeding in Vertebrates*, edited by J M Starck and T Wang, 423. New Hampshire, USA: Science Publishers Inc.

Andrade, D V, L F De Toledo, A S Abe, and T Wang. 2004. Ventilatory compensation of the alkaline tide during digestion in the snake boa constrictor. *The Journal of Experimental Biology* 207 (Pt 8): 1379–85. doi: 10.1242/jeb.00896.

Arvedsen, S K, J B Andersen, M Zaar, D Andrade, A S Abe, and T Wang. 2005. Arterial acid–base status during digestion and following vascular infusion of $NaHCO_3$ and HCl in the South American rattlesnake, *Crotalus durissus. Comparative Biochemistry and Physiology Part A: Molecular & Integrative Physiology* 142 (4): 495–502. doi: 10.1016/j.cbpa.2005.10.001.

Axelsson, M, H Thorarensen, S Nilsson, and A P Farrell. 2000. Gastrointestinal blood flow in the red Irish lord, *Hemilepidotus hemilepidotus*: Long-term effects of feeding and adrenergic control. *Journal of Comparative Physiology. B: Biochemical, Systemic, and Environmental Physiology* 170 (2): 145–52.

Bessler, S M, M C Stubblefield, G R Ultsch, and S M Secor. 2010. Determinants and modeling of specific dynamic action for the common garter snake (*Thamnophis sirtalis*). *Canadian Journal of Zoology/Revue Canadienne De Zoologie* 88 (8): 808–20. doi: 10.1139/Z10-045.

Bicego, K C, R C H Barros, and L G S Branco. 2007. Physiology of temperature regulation: Comparative aspects. *Comparative Biochemistry and Physiology Part A: Molecular & Integrative Physiology* 147 (3): 616–39. doi: 10.1016/j.cbpa.2006.06.032.

Bovo, R P, A Fuga, M A de Micheli-Campbell, J E Carvalho, and D V Andrade. 2014. Blood oxygen affinity increases during digestion in the South American rattlesnake, *Crotalus durissus terrificus*. *Comparative Biochemistry and Physiology Part A: Molecular & Integrative Physiology*, November 1–8. Elsevier B.V., doi: 10.1016/j.cbpa.2014.10.010.

Brett, J R. 1971. Energetic responses of salmon to temperature. A study of some thermal relations in the physiology and freshwater ecology of sockeye salmon (*Oncorhynchus nerka*). *American Zoologist* 11: 99–113.

Brown, C R and J N Cameron. 1991a. The induction of specific dynamic action in channel catfish by infusion of essential amino-acids. *Multiple Values Selected* 64 (1): 276–97.

Brown, C R and J N Cameron. 1991b. The relationship between specific dynamic action (SDA) and protein-synthesis rates in the channel catfish. *Physiological Zoology* 64 (1): 298–309.

Buddington, R K and J M Diamond. 1989. Ontogenetic development of intestinal nutrient transporters. *Annual Review of Physiology* 51: 601–19. doi: 10.1146/annurev.ph.51.030189.003125.

Busk, M, F B Jensen, and T Wang. 2000a. Effects of feeding on metabolism, gas transport, and acid–base balance in the bullfrog *Rana catesbeiana*. *American Journal of Physiology: Regulatory, Integrative and Comparative Physiology* 278 (1): R185–95.

Busk, M, J Overgaard, J W Hicks, A F Bennett, and T Wang. 2000b. Effects of feeding on arterial blood gases in the American alligator *Alligator mississippiensis*. *Journal of Experimental Biology* 203 (Pt 20): 3117–24.

Campen, R and J M Starck. 2012. Physiological responses to starvation in snakes: Low energy specialists. In *Comparative Physiology of Fasting, Starvation, and Food Limitation*, edited by M D McCue, 133–54. Berlin, Heidelberg: Springer-Verlag. doi: 10.1007/978-3-642-29056-5_8.

Canjani, C, D V Andrade, A P Cruz-Neto, and A S Abe. 2002. Aerobic metabolism during predation by a boid snake. *Comparative Biochemistry and Physiology Part A: Molecular & Integrative Physiology* 133 (3): 487–98. Elsevier.

Casewell, N R, G A Huttley, and W Wüster. 2012. Dynamic evolution of venom proteins in squamate reptiles. *Nature Communications* 3 (September): 1066–1075. Nature Publishing Group. doi: 10.1038/ncomms2065.

Caviedes-Vidal, E, W H Karasov, J G Chediack, V Fasulo, A P Cruz-Neto, and L Otani. 2008. Paracellular absorption: A bat breaks the mammal paradigm. *PLoS ONE* 3 (1): e1425. Edited by S Humphries. doi: 10.1371/journal.pone.0001425.t002.

Chu, C-W, T-S Tsai, I-H Tsai, Y-S Lin, and M-C Tu. 2009. Prey envenomation does not improve digestive performance in Taiwanese pit vipers (*Trimeresurus gracilis* and *T. stejnegeri stejnegeri*). *Comparative Biochemistry and Physiology Part A: Molecular & Integrative Physiology* 152 (4): 579–85. doi: 10.1016/j.cbpa.2009.01.006.

Conlon, J M, T Adrian, and S Secor. 1997a. Tachykinins (substance P, neurokinin a and neuropeptide Γ) and neurotensin from the intestine of the Burmese python, *Python molurus*. *Peptides* 18 (10): 1505–10.

Conlon, J M, S M Secor, T E Adrian, D C Mynarcik, and J Whittaker. 1997b. Purification and characterization of islet hormones (insulin, glucagon, pancreatic, polypeptide and somatostatin) from the Burmese python, *Python molurus*. *Regulatory Peptides* 71 (3): 191–98. doi: 10.1016/S0167-0115(97)01030-6.

Coulson, R A and T Hernandez. 1979. Increase in metabolic rate of the alligator fed proteins or amino acids. *The Journal of Nutrition* 109 (4): 538–50.

Cox, C L and S M Secor. 2007. Effects of meal size, clutch, and metabolism on the energy efficiencies of juvenile Burmese pythons, *Python molurus*. *Comparative Biochemistry and Physiology Part A: Molecular & Integrative Physiology* 148 (4): 861–68. doi: 10.1016/j.cbpa.2007.08.029.

Cox, C L and S M Secor. 2008. Matched regulation of gastrointestinal performance in the Burmese python, *Python molurus*. *Journal of Experimental Biology* 211 (7): 1131–40. doi: 10.1242/jeb.015313.

Cox, H S, D M Kaye, J M Thompson, A G Turner, G L Jennings, C Itsiopoulos, and M D Esler. 1995. Regional sympathetic nervous activation after a large meal in humans. *Clinical Science* 89 (2): 145–54.

Cruz-Neto, A P, D V Andrade, and A S Abe. 1999. Energetic cost of predation: Aerobic metabolism during prey ingestion by juvenile rattlesnakes, *Crotalus durissus*. *Journal of Herpetology* 33 (2): 229–34.

Cundall, D, C Tuttman, and M Close. 2014. A model of the anterior esophagus in snakes, with functional and developmental implications. *Anatomical Record* 297 (3): 586–98. Edited by J. D Daza and S C Miller. doi: 10.1002/ar.22860.

Dawson, W R. 1975. On the physiological significance of the preferred body temperatures of reptiles. *Perspectives of Biophysical Ecology* 12: 443–73. Ecological Studies. Berlin, Heidelberg: Springer Berlin Heidelberg. doi: 10.1007/978-3-642-87810-7_25.

Diefenbach, C. 1975. Gastric function in *Caiman crocodilus* (Crocodylia: Reptilia)—I. Rate of gastric digestion and gastric motility as a function of temperature. *Comparative Biochemistry and Physiology Part A: Physiology* 51 (2): 259–65. doi: 10.1016/0300-9629(75)90369-2.

Eliason, E J, D A Higgs, and A P Farrell. 2008. Postprandial gastrointestinal blood flow, oxygen consumption and heart rate in rainbow trout (*Oncorhynchus mykiss*). *Comparative Biochemistry and Physiology Part A: Molecular & Integrative Physiology* 149 (4): 380–88. doi: 10.1016/j.cbpa.2008.01.033.

Enok, S, L S Simonsen, S V Pedersen, T Wang, and N Skovgaard. 2012. Humoral regulation of heart rate during digestion in pythons (*Python molurus* and *Python regius*). *AJP: Regulatory, Integrative and Comparative Physiology* 302 (March): R1176–83. doi: 10.1152/ajpregu.00661.2011.

Enok, S, L S Simonsen, and T Wang. 2013. The contribution of gastric digestion and ingestion of amino acids on the postprandial rise in oxygen consumption, heart rate and growth of visceral organs in pythons. *Comparative Biochemistry and Physiology Part A: Molecular & Integrative Physiology* 165 (1): 46–53. doi: 10.1016/j.cbpa.2013.01.022.

Ferraris, R P, S A Villenas, B A Hirayama, and J Diamond. 1992. Effect of diet on glucose transporter site density along the intestinal crypt-villus axis. *The American Journal of Physiology* 262 (6 Pt 1): G1060–68.

Fry, B G, N Vidal, J A Norman, F J Vonk, H Scheib, S F Ryan Ramjan, S Kuruppu et al. 2005. Early evolution of the venom system in lizards and snakes. *Nature* 439 (7076): 584–88. doi: 10.1038/nature04328.

Fry, F E J. 1971. The effect of environmental factors on the physiology of fish. *Fish Physiology* 6: 1–98. Elsevier. doi: 10.1016/S1546-5098(08)60146-6.

Garland, T and S J Arnold. 1983. Effects of a full stomach on locomotory performance of juvenile garter snakes (*Thamnophis elegans*). *Copeia* 1983: 1092–96.

Geiser, F. 2004. Metabolic rate and body temperature reduction during hibernation and daily torpor. *Annual Review of Physiology* 66: 239–74. doi: 10.1146/annurev.physiol.66.032102.115105.

Geiser, F and C Stawski. 2011. Hibernation and torpor in tropical and subtropical bats in relation to energetics, extinctions, and the evolution of endothermy. *Integrative and Comparative Biology* 51: 337–348.

Greene, H W. 1997. *Snakes—The Evolution of Mystery in Nature*. Berkely, Los Angeles, London: University of California Press.

Guppy, M and P Withers. 1999. Metabolic depression in animals: Physiological perspectives and biochemical generalizations. *Biological Reviews of the Cambridge Philosophical Society* 74 (1): 1–40.

Hagey, L R, P R Møller, A F Hofmann, and M D Krasowski. 2010a. Diversity of bile salts in fish and amphibians: Evolution of a complex biochemical pathway. *Physiological and Biochemical Zoology: PBZ* 83 (2): 308–21. doi: 10.1086/649966.

Hagey, L R, N Vidal, A F Hofmann, and M D Krasowski. 2010b. Evolutionary diversity of bile salts in reptiles and mammals, including analysis of ancient human and extinct giant ground sloth coprolites. *BMC Evolutionary Biology* 10: 133. doi: 10.1186/1471-2148-10-133.

Hansen, K, Pedersen, PB, Pedersen, M, and Wang, T. 2013. Magnetic resonance imaging volumetry for noninvasive measures of phenotypic flexibility during digestion in Burmese pythons. *Physiological and Biochemical Zoology*. 86 (1): 149–58.

Harlow, H J, S S Hillman, and M Hoffman. 1976. The effect of temperature on digestive efficiency in the herbivorous lizard, *Dipsosaurus dorsalis*. *Journal of Comparative Physiology B: Biochemical, Systemic, and Environmental Physiology* 111 (1): 1–6. doi: 10.1007/BF00691105.

Haslewood, G A. 1967. Bile salt evolution. *Journal of Lipid Research* 8 (6): 535–50.

Helmstetter, C, R K Pope, M T'Flachebba, S M Secor, and J-H Lignot. 2009a. The effects of feeding on cell morphology and proliferation of the gastrointestinal tract of juvenile Burmese pythons (*Python molurus*). *Canadian Journal of Zoology/Revue Canadienne De Zoologie* 87 (12): 1255–67. doi: 10.1139/Z09-110.

Helmstetter, C, N Reix, M T'Flachebba, R K Pope, S M Secor, Y L Maho, and J-H Lignot. 2009b. Functional changes with feeding in the gastro-intestinal epithelia of the Burmese python (*Python molurus*). *Zoological Science* 26 (9): 632–38. doi: 10.2108/zsj.26.632.

Henriksen, PS, S Enok, J Overgaard, and T Wang. 2015. Food composition influences metabolism, heart rate and organ growth during digestion in Python regius. *Comparative Biochemistry and Physiology Part A: Molecular & Integrative Physiology*. 183: 36–44.

Hicks, J W, T Wang, and A F Bennett. 2000. Patterns of cardiovascular and ventilatory response to elevated metabolic states in the lizard *Varanus exanthematicus*. *The Journal of Experimental Biology* 203 (Pt 16): 2437–45.

Hofmann, A F, L R Hagey, and M D Krasowski. 2010. Bile salts of vertebrates: Structural variation and possible evolutionary significance. *Journal of Lipid Research* 51 (2): 226–46. doi: 10.1194/jlr.R000042.

Holst, J J, T Vilsbøll, and C F Deacon. 2009. The incretin system and its role in type 2 diabetes mellitus. *Molecular and Cellular Endocrinology* 297 (1–2): 127–36. doi: 10.1016/j.mce.2008.08.012.

Horn, M H and K S Messer. 1992. Fish guts as chemical reactors: A model of the alimentary canals of marine herbivorous fishes. *Marine Biology* 113: 527–35.

Huey, R B and P E Hertz. 1984. Is a jack-of-all-temperatures a master of none? *Evolution* 38(2): 441–44.

Huey, R B and R D Stevenson. 1979. Integrating thermal physiology and ecology of ectotherms: A discussion of approaches. *Integrative and Comparative Biology* 19 (1): 357–66. doi: 10.1093/icb/19.1.357.

Hume, I D. 1989. Optimal digestive strategies in mammalian herbivores. *Physiological Zoology* 62 (6): 1145–63. University of Chicago Press.

Humphries, M M, D W Thomas, and D L Kramer. 2003. The role of energy availability in mammalian hibernation: A cost-benefit approach. *Physiological and Biochemical Zoology: PBZ* 76 (2): 165–79. The University of Chicago Press. doi: 10.1086/367950.

Ingelfinger, F J. 1958. Esophageal motility. *Physiological Reviews* 38 (4): 533–84. American Physiological Society.

Jensen, B, C K Larsen, J M Nielsen, L S Simonsen, and T Wang. 2011. Change of cardiac function, but not form, in postprandial pythons. *Comparative Biochemistry and Physiology Part A: Molecular & Integrative Physiology* 160 (1): 35–42. doi: 10.1016/j.cbpa.2011.04.018.

Jobling, M. 1987. Influences of food particle size and dietary energy content on patterns of gastric evacuation in fish: Test of a physiological model of gastric emptying. *Journal of Fish Biology* 30 (3): 299–314. doi: 10.1111/j.1095-8649.1987. tb05754.x.

Jobling, M and P Spencer Davies. 1980. Effects of feeding on metabolic rate, and the specific dynamic action in plaice, *Pleuronectes platessa* L. *Journal of Fish Biology* 16 (6): 629–38. doi: 10.1111/j.1095-8649.1980.tb03742.x.

Johnson, L R. 1977. Gastrointestinal hormones and their functions. *Annual Review of Physiology* 39: 135–58. doi: 10.1146/annurev.ph.39.030177.001031.

Johnson, L R, G A Castro, E D Jacobsen, and N W Weisbrodt. 1991. *Gastrointestinal Physiology*. Edited by Johnson Leonard R. Mosby Elsevier Health Science, 176 pages (first published 1977), St. Louis.

Karasov, W H. 2011. Digestive physiology: A view from molecules to ecosystem. *AJP: Regulatory, Integrative and Comparative Physiology* 301 (2): R276–84. doi: 10.1152/ajpregu.00600.2010.

Karasov, W H and J M Diamond. 1983. Adaptive regulation of sugar and amino acid transport by vertebrate intestine. *The American Journal of Physiology* 245 (4): G443–62.

Karasov, W H and J M Diamond. 1988. Interplay between physiology and ecology in digestion. *Bioscience* 38: 602–11.

Karasov, W H and A E Douglas. 2013. Comparative digestive physiology. *Comprehensive Physiology* 3 (2): 741–83. doi: 10.1002/cphy.c110054.

Karasov, W H and I D Hume. 2010. *Vertebrate Gastrointestinal System. Comprehensive Physiology*. Hoboken, New Jersey: John Wiley & Sons, Inc. doi: 10.1002/cphy. cp130107.

Karasov, W H, C Martínez del Rio, and E Caviedes-Vidal. 2011. Ecological physiology of diet and digestive systems. *Annual Review of Physiology* 73 (1), 69–93. doi: 10.1146/annurev-physiol-012110-142152.

Karasov, W H, D H Solberg, and J M Diamond. 1987. Dependence of intestinal amino acid uptake on dietary protein or amino acid levels. *The American Journal of Physiology* 252 (5 Pt 1): G614–25.

Leite, C A C, T Wang, E W Taylor, A S Abe, G S P C Leite, and D O V de Andrade. 2014. Loss of the ability to control right-to-left shunt does not influence the metabolic responses to temperature change or long-term fasting in the South American rattlesnake *Crotalus durissus*. *Physiological and Biochemical Zoology: PBZ* 87 (4): 568–75. doi: 10.1086/675863.

Lignot, J-H, C Helmstetter, and S M Secor. 2005. Postprandial morphological response of the intestinal epithelium of the Burmese python (*Python molurus*). *Comparative Biochemistry and Physiology Part A: Comparative Physiology* 141 (3): 280–91. doi: 10.1016/j.cbpb.2005.05.005.

Lillywhite, H, P. de Delva, and B P Noonan. 2002. Patterns of gut passage time and the chronic retention of fecal mass in viperid snakes. In G W Schuett, M Höggren, and H W Greene (eds.), *Biology of the Vipers*, pp. 497–506, Traverse City, MI: Biological Sciences Press.

Lin, H C, C Neevel, and J H Chen. 2004. Slowing intestinal transit by PYY depends on serotonergic and opioid pathways. *American Journal of Physiology – Gastrointestinal and Liver Physiology* 286 (4): G558–63. doi: 10.1152/ajpgi.00278.2003.

Low, A G. 2009. Nutritional regulation of gastric secretion, digestion and emptying. *Nutrition Research Reviews* 3 (01): 229. doi: 10.1079/NRR19900014.

Lusk, G. 1922. The specific dynamic action of various food factors. *Medicine* 1 (2): 311.

Madsen, J L, S B Søndergaard, and S Møller. 2006. Meal-induced changes in splanchnic blood flow and oxygen uptake in middle-aged healthy humans. *Scandinavian Journal of Gastroenterology* 41 (1): 87–92. doi: 10.1080/00365520 510023882.

Marieb, E and K Hoehn. 2007. *Human Anatomy & Physiology*. San Francisco: Pearson Benjamin Cummings.

McCue, M. 2005. Enzyme activities and biological functions of snake venoms. *Applied Herpetology* 2 (2): 109–23. doi: 10.1163/1570754043492135.

McCue, M D. 2006. Specific dynamic action: A century of investigation. *Comparative Biochemistry and Physiology Part A: Comparative Physiology* 144: 381–94. doi: 10.1016/j.cbpa.2006.03.011.

McCue, M D. 2007a. Prey envenomation does not improve digestive performance in western diamondback rattlesnakes (*Crotalus atrox*). *Journal of Experimental Zoology Part A: Comparative Experimental Biology* 307 (10): 568–77. doi: 10.1002/jez.411.

McCue, M D. 2007b. Snakes survive starvation by employing supply-and demand-side economic strategies. *Zoology* 110 (4): 318–27. doi: 10.1016/j.zool.2007.02.004.

McCue, M D, A F Bennett, and J W Hicks. 2005. The effect of meal composition on specific dynamic action in Burmese pythons (*Python molurus*). *Physiological and Biochemical Zoology: PBZ* 78 (2): 182–92. doi: 10.1086/427049.

McCue, M D and H B Lillywhite. 2002. Oxygen consumption and the energetics of island-dwelling Florida cottonmouth snakes. *Physiological and Biochemical Zoology: PBZ* 75 (2): 165–78. doi: 10.1086/339390.

McKinstry, D M. 1978. Evidence of toxic saliva in some colubrid snakes of the United States. *Toxicon* 16 (6): 523–34. doi: 10.1016/0041-0101(78)90179-4.

McWhorter, T J, B Pinshow, W H Karasov, and C R Tracy. 2013. Paracellular absorption is relatively low in the herbivorous Egyptian spiny-tailed lizard, *Uromastyx aegyptia. PLoS ONE* 8 (4): e61869. doi: 10.1371/journal.pone.0061869. s001.

Milsom, W K, D V Andrade, S P Brito, L F Toledo, T Wang, and A S Abe. 2008. Seasonal changes in daily metabolic patterns of tegu lizards (*Tupinambis merianae*) placed in the cold (17 degrees C) and dark. *Physiological and Biochemical Zoology: PBZ* 81 (2): 165–75. doi: 10.1086/524148.

Moschetta, A, F Xu, L R Hagey, G P van Berge-Henegouwen, K J van Erpecum, J F Brouwers, J C Cohen et al. 2005. A phylogenetic survey of biliary lipids in vertebrates. *Journal of Lipid Research* 46 (10): 2221–32. doi: 10.1194/jlr. M500178-JLR200.

Moscona, A A. 1990. Anatomy of the pancreas and Langerhans islets in snakes and lizards. *Anatomical Record* 227 (2): 232–44. doi: 10.1002/ar.1092270212.

Murphy, K G and S R Bloom. 2006. Gut hormones and the regulation of energy homeostasis. *Nature* 444 (7121): 854–59. doi: 10.1038/nature05484.

Navas, C A and J E Carvalho (Eds.). 2010. *Aestivation: Molecular and Physiological Aspects.* Berlin, Heidelberg: Springer-Verlag.

Nørgaard, S, K Andreassen, C L Malte, and S Enok, and T Wang. 2016. Low cost of gastric acid secretion during digestion in ball pythons, *Comparative Biochemistry and Physiology Part A: Molecular & Integrative Physiology* 194: 62–66.

Ott, B D and S M Secor. 2007. Adaptive regulation of digestive performance in the genus *Python. The Journal of Experimental Biology* 210 (Pt 2): 340–56. doi: 10.1242/jeb.02626.

Overgaard, J, J B Andersen, and T Wang. 2002. The effects of fasting duration on the metabolic response to feeding in *Python molurus*: An evaluation of the energetic costs associated with gastrointestinal growth and upregulation. *Physiological and Biochemical Zoology: PBZ* 75 (4): 360–68. doi: 10.1086/342769.

Overgaard, J, M Busk, J W Hicks, F B Jensen, and T Wang. 1999. Respiratory consequences of feeding in the snake *Python molorus. Comparative Biochemistry and Physiology Part A: Molecular & Integrative Physiology* 124 (3): 359–65. doi: 10.1016/S1095-6433(99)00127-0.

Overgaard, J and T Wang. 2002. Increased blood oxygen affinity during digestion in the snake *Python molurus. Journal of Experimental Biology* 205 (Pt 21): 3327–34.

Pappenheimer, J R. 1993. On the coupling of membrane digestion with intestinal absorption of sugars and amino acids. *The American Journal of Physiology* 265 (3 Pt 1): G409–17.

Pappenheimer, J R and K Z Reiss. 1987. Contribution of solvent drag through intercellular junctions to absorption of nutrients by the small intestine of the rat. *The Journal of Membrane Biology* 100 (2): 123–36.

Pennisi, E. 2005. The dynamic gut. *Science (New York, N.Y.).* 307(5717): 1896–1899. doi: 10.1126/science.307.5717.1896.

Perko, M J, H B Nielsen, C Skak, J O Clemmesen, T V Schroeder, and N H Scher. 1998. Mesenteric, coeliac and splanchnic blood flow in humans during exercise. *The Journal of Physiology* 513 (3): 907–13.

Piersma, T and A Lindström. 1997. Rapid reversible changes in organ size as a component of adaptive behaviour. *Trends in Ecology and Evolution* 12 (4): 134–38.

Price, E R, A Brun, E Caviedes-Vidal, and W H Karasov. 2015. Digestive adaptations of aerial lifestyles. *Physiology (Bethesda, MD)* 30 (1): 69–78. doi: 10.1152/physiol.00020.2014.

Rees, W D W, J R Malagelada, L J Miller, and V L W Go. 1982. Human interdigestive and postprandial gastrointestinal motor and gastrointestinal hormone patterns. *Digestive Diseases and Sciences* 27 (4): 321–29. doi: 10.1007/BF01296751.

Rindi, G, A B Leiter, A S Kopin, C Bordi, and E Solcia. 2004. The "normal" endocrine cell of the gut: Changing concepts and new evidences. *Annals of the New York Academy of Sciences* 1014 (1): 1–12. doi: 10.1196/annals.1294.001.

Riquelme, C A, J A Magida, B C Harrison, C E Wall, T G Marr, S M Secor, and L A Leinwand. 2011. Fatty acids identified in the Burmese python promote beneficial cardiac growth. *Science (New York, NY)* 334 (6055): 528–31. doi: 10.1126/science.1210558.

Rubner, M. 1902. *Die Gesetze des Energieverbrauchs bei der Ernährung*. Leipzig, Wien: Deuticke, 426 pp.

Sanford, P A. 1992. *Digestive System Physiology*, 2nd edn. London: Arnold, 251 pp.

Secor, S. 1995. Digestive response to the first meal in hatchling Burmese pythons (*Python molurus*). *Copeia* 4: 947–54.

Secor, S. 1997. Determinants of the postfeeding metabolic response of Burmese pythons, *Python molurus*. *Physiological Zoology* 70: 202–12.

Secor, S, J Lane, and E Whang. 2002. Luminal nutrient signals for intestinal adaptation in pythons. *American Journal of Physiology – Gastrointestinal and Liver Physiology* 282: G1298–309. doi: 10.1152/ajpgi.00194.2002.

Secor, S M. 2003. Gastric function and its contribution to the postprandial metabolic response of the Burmese python *Python molurus*. *Journal of Experimental Biology* 206 (Pt 10): 1621–30.

Secor, S M. 2005. Evolutionary and cellular mechanisms regulating intestinal performance of amphibians and reptiles. *American Zoologist* 45 (2): 282–94. doi: 10.1093/icb/45.2.282.

Secor, S M. 2008. Digestive physiology of the Burmese python: Broad regulation of integrated performance. *The Journal of Experimental Biology* 211 (Pt 24): 3767–74. doi: 10.1242/jeb.023754.

Secor, S M. 2009. Specific dynamic action: A review of the postprandial metabolic response. *Journal of Comparative Physiology B: Biochemical, Systemic, and Environmental Physiology* 179 (1): 1–56. doi: 10.1007/s00360-008-0283-7.

Secor, S M and J Diamond. 1995. Adaptive responses to feeding in Burmese pythons: Pay before pumping. *Journal of Experimental Biology* 198 (Pt 6): 1313–25.

Secor, S M and J Diamond. 1997. Effects of meal size on postprandial responses in juvenile Burmese pythons (*Python molurus*). *The American Journal of Physiology* 272 (3 Pt 2): R902–12.

Secor, S M and J Diamond. 1998. A vertebrate model of extreme physiological regulation. *Nature* 395 (6703): 659–62. doi: 10.1038/27131.

Secor, S M and J M Diamond. 2000. Evolution of regulatory responses to feeding in snakes. *Physiological and Biochemical Zoology: PBZ* 73 (2): 123–41. doi: 10.1086/316734.

Secor, S M and S E White. 2010. Prioritizing blood flow: Cardiovascular performance in response to the competing demands of locomotion and digestion for the Burmese python, Python molurus. *The Journal of Experimental Biology* 213: 78–88.

Secor, S M, D Fehsenfeld, J Diamond, and T E Adrian. 2001. Responses of python gastrointestinal regulatory peptides to feeding. *Proceedings of the National Academy of Sciences of the United States of America* 98 (24): 13637–42. doi: 10.1073/pnas.241524698.

Secor, S M, J W Hicks, and A F Bennett. 2000a. Ventilatory and cardiovascular responses of a python (*Python molurus*) to exercise and digestion. *The Journal of Experimental Biology* 203 (Pt 16): 2447–54.

Secor, S M, E D Stein, and J Diamond. 1994. Rapid upregulation of snake intestine in response to feeding: A new model of intestinal adaptation. *The American Journal of Physiology* 266 (4 Pt 1): G695–705.

Secor, S M, J R Taylor, and M Grosell. 2012. Selected regulation of gastrointestinal acid–base secretion and tissue metabolism for the diamondback water snake and Burmese python. *The Journal of Experimental Biology* 215 (Pt 1): 185–96. doi: 10.1242/jeb.056218.

Secor, S M, E E Whang, J S Lane, S W Ashley, and J Diamond. 2000b. Luminal and systemic signals trigger intestinal adaptation in the juvenile python. *American Journal of Physiology – Gastrointestinal and Liver Physiology* 279 (6): G1177–87.

Seth, H, E Sandblom, and M Axelsson. 2009. Nutrient-induced gastrointestinal hyperemia and specific dynamic action in rainbow trout (*Oncorhynchus mykiss*)—importance of proteins and lipids. *American Journal of Physiology: Regulatory, Integrative and Comparative Physiology* 296 (2): R345–52. doi: 10.1152/ajpregu.90571.2008.

Simpson, K A, N M Martin, and S R Bloom. 2008. Hypothalamic regulation of appetite. *Expert Review of Endocrinology & Metabolism* 3 (5): 577–92. doi: 10.1586/17446651.3.5.577.

Skovgaard, N, J M Conlon, and T Wang. 2007. Evidence that neurotensin mediates postprandial intestinal hyperemia in the python, *Python regius. American Journal of Physiology: Regulatory, Integrative and Comparative Physiology* 293 (3): R1393–99. doi: 10.1152/ajpregu.00256.2007.

Skovgaard, N, G Galli, E W Taylor, J M Conlon, and T Wang. 2005. Hemodynamic effects of python neuropeptide [gamma] in the anesthetized python, *Python regius. Regulatory Peptides* 128 (1): 15–26. doi: 10.1016/j.regpep.2004.12.016.

Skovgaard, N, K Møller, H Gesser, and T Wang. 2009. Histamine induces postprandial tachycardia through a direct effect on cardiac H-2-receptors in pythons. *American Journal of Physiology: Regulatory, Integrative and Comparative Physiology* 296 (3): R774–85. doi: 10.1152/ajpregu.90466.2008.

Slay, C E, S Enok, J W Hicks, and T Wang. 2014. Reduction of blood oxygen levels enhances postprandial cardiac hypertrophy in Burmese python (*Python bivittatus*). *The Journal of Experimental Biology* 217 (10): 1784–89. doi: 10.1242/jeb.092841.

Starck, J M. 2002. Structural flexibility of the small intestine and liver of garter snakes in response to feeding and fasting. *Journal of Experimental Biology* 205: 1377–88.

Starck, J M and K Beese. 2001. Structural flexibility of the intestine of Burmese python in response to feeding. *The Journal of Experimental Biology* 204 (Pt 2): 325–35.

Starck, J M, A P Cruz-Neto, and A S Abe. 2007. Physiological and morphological responses to feeding in broad-nosed Caiman (*Caiman latirostris*). *Journal of Experimental Biology* 210 (Pt 12): 2033–45. doi: 10.1242/jeb.000976.

Starck, J M and T Wang, eds. 2005. *Physiological and Ecological Adaptations to Feeding in Vertebrates*. New Hampshire, USA: Science Publishers Inc. http://scholar.google.com/scholar?q=related:1gP961B3upUJ:scholar.google. com/&hl=en&num=30&as_sdt=0,5.

Starck, J M and C Wimmer. 2005. Patterns of blood flow during the postprandial response in ball pythons, *Python regius*. *Journal of Experimental Biology* 208 (Pt 5): 881–89. doi: 10.1242/jeb.01478.

Stevens, C E and I D Hume. 2004. *Comparative Physiology of the Vertebrate Digestive System*. Cambridge: Cambridge University Press.

Taylor, E W, D Jordan, and J H Coote. 1999. Central control of the cardiovascular and respiratory systems and their interactions in vertebrates. *Physiological Reviews* 79 (3): 855–916.

Thomas, R G and F H Pough. 1979. The effect of rattlesnake venom on digestion of prey. *Toxicon* 17 (3): 221–28.

Toledo, L F, A S Abe, and D V Andrade. 2003. Temperature and meal size effects on the postprandial metabolism and energetics in a boid snake. *Physiological and Biochemical Zoology: PBZ* 76 (2): 240–46. doi: 10.1086/374300.

Toledo, L F, S P Brito, W K Milsom, A S Abe, and D V Andrade. 2008. Effects of season, temperature, and body mass on the standard metabolic rate of tegu lizards (*Tupinambis merianae*). *Physiological and Biochemical Zoology: PBZ* 81 (2): 158–64. doi: 10.1086/524147.

Tsai, T-S, H-J Lee, and M-C Tu. 2008. Specific dynamic action, apparent assimilation efficiency, and digestive rate in an arboreal pitviper, *Trimeresurus stejnegeri stejnegeri*. *Canadian Journal of Zoology-Revue Canadienne De Zoologie* 86 (10): 1139–51. doi: 10.1139/Z08-090.

Uriona, T J. 2005. Structure and function of the esophagus of the American alligator (*Alligator mississippiensis*). *The Journal of Experimental Biology* 208 (16): 3047–53. doi: 10.1242/jeb.01746.

Van Denburgh, J. 1898. Some experiments with the saliva of the Gila monster (*Heloderma suspectum*). *Transactions of the American Philosophical Society* 19 (2): 199. doi: 10.2307/1005452.

Waas, S, R A Werner, and J M Starck. 2010. Fuel switching and energy partitioning during the postprandial metabolic response in the ball python (*Python regius*). *Journal of Experimental Biology* 213 (8): 1266–71. doi: 10.1242/jeb.033662.

Wang, T. 2001. Introduction to symposium: The physiological effects of feeding. *Comparative Biochemistry and Physiology* 128A, 395–96.

Wang, T, M Axelsson, J Jensen, and J M Conlon. 2000. Cardiovascular actions of python bradykinin and substance P in the anesthetized python, *Python regius*. *American Journal of Physiology: Regulatory, Integrative and Comparative Physiology* 279 (2): R531–38.

Wang, T, W Burggren, and E Nobrega. 1995. Metabolic, ventilatory, and acid–base responses associated with specific dynamic action in the toad *Bufo marinus*. *Physiological Zoology* 68: 192–205. http://www.jstor.org/ stable/10.2307/30166499.

Wang, T, M Busk, and J Overgaard. 2001a. The respiratory consequences of feeding in amphibians and reptiles. *Comparative Biochemistry and Physiology Part A: Molecular & Integrative Physiology* 128 (3): 535–49.

Wang, T, C C Y Hung, and D J Randall. 2006. The comparative physiology of food deprivation: From feast to famine. *Annual Reviews* 68: 223–51. doi: 10.1146/ annurev.physiol.68.040104.105739.

Wang, T and N Skovgaard. 2008. The cardio-respiratory response to increased metabolic rate during digestion. *Comparative Biochemistry and Physiology Part A: Molecular & Integrative Physiology* 150 (3): S62–S62. doi: 10.1016/j.cbpa.2008.04.066.

Wang, T, E Taylor, and D Andrade. 2001b. Autonomic control of heart rate during forced activity and digestion in the snake boa constrictor. *The Journal of Experimental Biology* 204: 3553–60.

Wang, T, M Zaar, S Arvedsen, C Vedel-Smith, and J Overgaard. 2002. Effects of temperature on the metabolic response to feeding in *Python molurus*. *Comparative Biochemistry and Physiology Part A—Molecular & Integrative Physiology* 133: 519–27.

chapter five

Effects of feeding on the respiration of ectothermic vertebrates

José Eduardo de Carvalho, Denis Vieira de Andrade,
and William K. Milsom

Contents

Introduction

Many ectothermic vertebrates are opportunistic carnivores that ingest large meals at irregular intervals (Greene 1997, Secor and Diamond 1998, 2000, Secor 2001, Secor and Boehm 2006, Cocker-Butar and Secor 2014). Interspecific differences in feeding habits give rise to specific adjustments in a number of physiological processes directly or indirectly involved in the digestion and absorption of nutrients (Secor 2001, Wang 2001, Wang et al. 2001, McCue 2006, Secor 2009, and see also Chapter 4 in this book). The ingestion of large meals in some ectotherms is followed by an steep rise in the secretion of digestive fluids and in the capacity for intestinal assimilation and, in some cases, by changes in the relative size of the organs involved in the digestive processes (Wang et al. 2001, Secor 2009). Associated with such events, there is a considerable increase in the energy requirement during the postprandial period leading to cardiorespiratory adjustments to ensure an adequate supply of oxygen (Wang et al. 2001, Andersen et al. 2005, McCue 2006). Despite the fundamental importance of these adjustments to match the increased energy demands associated with meal digestion, while correcting ionic and acid/base disturbances, very little is known about the mechanisms that coordinate the cardiovascular and respiratory changes during the postprandial period. In this chapter, we will first review the nature of the changes in energy demand associated with feeding in different groups of ectotherms and then examine what is known about the cardiorespiratory changes associated with them.

Energetic correlates of feeding in ectothermic vertebrates

In general, aerobic capacity is correlated with maximum athletic performance as documented in mammals (Seeherman et al. 1981, Brooks 1998, Weibel et al. 2004); birds (Suarez 1996, 1998); lizards (Bennett 1994); amphibians (Taigen and Wells 1985, Navas et al. 2008); fishes (Owen 2001); and also invertebrates (Suarez 1998). However, for many species of ectotherms, the energy cost associated with the digestive and absorptive processes can exceed the cost of the muscular activity associated with locomotion (Secor and Diamond 1995, 1997, Andrade et al. 1997, Secor et al. 2000, Hicks and Bennett 2004). This is especially true for the ectothermic species that are opportunistic carnivores and are able to ingest large meals at irregular intervals.

The increase in metabolic rate associated with the ingestion of large meals has been termed as "specific dynamic action" (or SDA; Kleiber 1961, and see Chapter 4 in this book). Most of this postprandial metabolic increase represents the sum of the energy costs involved in the ingestion, digestion, absorption, and assimilation of nutrients. The magnitude of the SDA is influenced by factors such as the size of the meal and its composition (Secor 2003, McCue 2006, Secor et al. 2007, Secor 2009). In the boid snake *Boa constrictor*, for example, the maximum values of oxygen consumption and the duration of the postprandial period are related to the size of the prey. Increases in body temperature induced by environmental temperature significantly reduce the postprandial period while increasing peak rates of oxygen consumption (Toledo et al. 2003). The total amount of energy required for the process, however, is not greatly affected by the body temperature; then, the relative size of the prey is the most important factor determining the magnitude of the SDA in boas (see Wang et al. 2005, McCue 2006, Luo and Xie 2008, Secor 2009).

The question arises as to what happens if digestion and exercise occur simultaneously. On the basis of studies carried out with fish, Owen (2001) proposed that the maximum oxygen consumption possible is the sum of the energy expended on digestion and absorption (i.e., SDA) and that spent on other activities such as locomotion. More energy spent on one process was at the expense of energy available for other processes. This would explain why maximum swimming speeds were lower in fed fishes relative to fish in the postabsorptive period (Owen 2001, Jordan and

Steffensen 2005, Wang et al. 2012). Hicks and Bennett (2004), however, found that the aerobic capacity of varanid lizards submitted to intense running exercise in the postprandial period was elevated when compared to that of fasting animals. This observation indicates that concurrent activities, such as locomotion and digestion, may additively contribute to the overall increment in oxygen consumption of the whole animal (Hicks and Bennett 2004, but see also McCue 2006). The extent to which the ability to separate the energy expenditure associated with the digestive process and that associated with other activities which varies interspecifically among ectothermic vertebrates remains to be explored.

Given that energy expenditure during the postprandial period of many ectotherms is the sum of the costs of different processes (from prey subjugation to the completion of nutrient absorption and waste disposal) accounting for different portions of the total investment and occurring at different times, we find temporal differences in the requirements for oxygen and substrate delivery to the different tissues and organs. Studies of Secor and Diamond (2000) clearly demonstrated that there are only modest metabolic responses in snakes that frequently consume small prey compared with snakes that episodically feed on large prey after long periods of fasting. In the species of snakes that feed infrequently, the tissues of the gastrointestinal tract are subject to atrophy during the intervals between feedings, a strategy suggested to provide energy savings on the cost of maintenance (Beaupre and Duvall 1998, Secor 2001, but see Starck and Beese 2002, Waas et al. 2010). Starck and Beese (2001, 2002) observed that the regulation of the size of the enteric and hepatic tissues after feeding in two species of snakes (*Thamnophis sirtalis* and *Python bivittatus*) involved only the osmotically driven transport of fluids from blood and lymph to the inactive cells as well as the incorporation of fat droplets. There was no evidence of cell proliferation during the process of tissue rearrangement after meal ingestion (Starck and Beese 2002). These events occur rapidly, increasing, at no cost, the size of these tissues to threefold in relation to the fasting period (Starck and Beese 2001, 2002). The acute increase in metabolism following the ingestion of prey by these snakes must have arisen from something other than the morphological restructuring of the gastrointestinal tract and accessory organs (see Overgaard et al. 2002, Holmberg et al. 2003, Secor 2003, Lignot et al. 2005, Henriksen et al. 2015, McCue et al. 2015). It is estimated that this is in part due to changes in the activity of carbohydrate, amino acid, and other carriers in the gastrointestinal epithelium (Secor and Diamond 2000, Secor et al. 2001, Iglesias et al. 2003, but see also Secor et al. 2007). Secor (2001), working with *P. bivittatus*, demonstrated that the synthesis and secretion of hydrochloric acid (HCl) and pepsinogen accounted for most of the energy expenditure during digestion (more than 55% of SDA; see also Secor and Diamond 1995, 1998) while Henriksen et al. (2015), working with *Python regius*, demonstrated that protein synthesis was the major contributor to the SDA and that increased gastric acid secretion was not metabolically costly. Similar conclusions were drawn from boas whose gastric acid secretion was pharmacologically inhibited (Andrade et al., 2004a).

Given the intensity and duration of metabolic changes in the postprandial period (Preest 1994, Busk et al. 2000a, McCue 2006), it is conceivable that glycolytic pathways may play a secondary role in the maintenance of energy balance in tissues directly and indirectly associated with the digestion and absorption of nutrients. According to Kemper et al. (2001), anaerobic metabolism, resulting in lactate production, significantly helps to increase the overall capacity of the specialized shaker muscles in the tails of rattlesnakes to maintain energy balance during intense activity, even when oxygen availability is not limiting (see Brooks 1998, Conley et al. 1998, Brooks et al. 1999, Gladden 2001, Shulman and Rothman 2001, Hochachka and Somero 2002). Certainly, in *Alligator mississipiensis*, the increased aerobic metabolism was accompanied by a small, but significant, increase in plasma lactate concentration between 24 and 72 h following feeding (Busk et al. 2000b, also see Hartzler et al. 2006, e.g., in varanid lizards). Although the contribution in this instance was small, it indicates that the glycolytic pathway is recruited and could be important for the maintenance of energy balance during the postprandial period. It also suggests that a more complete understanding of the metabolic modifications that follow the ingestion of large prey in ectotherms with different feeding strategies will be required to reduce the risk of overgeneralization.

The remarkable morphological changes and the postprandial metabolic responses just described (Secor et al. 1994, 2001, Secor and Diamond 1995, Secor 2001, 2003, Starck and Beese 2002, Holmberg et al. 2003, Lignot et al. 2005, and also Chapter 4 of Enok et al. in this book) are also accompanied by cardiorespiratory adjustments that ensure adequate O_2 uptake and acid–base changes that optimize the catabolic and anabolic processes that follow food intake (Overgaard et al. 1999, Busk et al. 2000a,b, Hicks et al. 2000, Secor et al. 2000, 2001, Wang et al. 2001, Andersen et al. 2003, Andrade et al. 2004a, Bovo et al. 2015). The functional significance of respiratory and cardiovascular changes in different groups of ectotherms as well as the regulatory mechanisms, however, remains obscure (see McCue 2006, Henriksen et al. 2015). In many ectotherms, large meals impair mobility and pulmonary ventilation (Garland and Arnold 1983, Milsom 1988, Hicks et al. 2000, Hicks and Bennett 2004). In addition, the presence of food in the stomach leads to intense acid secretion requiring cardiovascular and respiratory adjustments to compensate for the effects of alkalization of the blood and the drastic changes in the ionic balance between body compartments (Wang et al. 2001). In the next section, we explore these cardiovascular and respiratory adjustments that occur in ectothermic vertebrates during the period following the ingestion of food.

Respiratory and circulatory consequences of food intake in ectotherms

In fishes, snakes, lizards, and amphibians, an increase in total ventilation due to increases in breathing frequency and tidal volume accompanies the increase in metabolic rate associated with SDA. This serves to increase O_2 delivery to the gas exchange surfaces (Secor et al. 2000, Bennett and Hicks 2001, Busk et al. 2000a, Owen 2001, Andersen and Wang 2003).

In parallel with the increase in ventilation, a remarkable increase in heart rate (f_H) is observed immediately after prey ingestion (see Wang et al. 2001, Clark et al. 2005). In lizards, at least in monitors (*Varanus*), the increase in f_H is positively correlated with metabolic rate (Hicks et al. 2000). In addition, in several species of reptiles, the increase in f_H is accompanied by increases in stroke volume (up to 50% of the values observed for animals during intense physical activity) (Secor et al. 2000, Andersen et al. 2005) and cardiac output (Wang et al. 1995, 2001, Hicks et al. 2000, Secor et al. 2000, Hicks and Bennett 2004, Jensen et al. 2011). Andersen et al. (2005) found that in pythons (*P. bivittatus*), there was a conspicuous ventricular hypertrophy 48 h after prey ingestion, associated with an increase in cardiac myosin messenger ribonucleic acid (mRNA) expression. This change was completely reversible when the animal fasted again. In the case of pythons, this was also associated with the incorporation of more contractile elements into the ventricular muscle fibers as a direct result of the increased transcription of cardiac genes (Andersen et al. 2005). This cardiac hypertrophy clearly contributes to the increased cardiac output during the postprandial period increasing the capacity for oxygen transport (see Chapter 4 in this book).

Also, it is important to remember that both reptiles and amphibians have an incomplete anatomical division of the ventricle and that the mixture of more oxygen-rich blood from the lungs with the venous blood from the systemic circulation determines the O_2 content of systemic arterial blood (Wood 1984, Wang and Hicks 1996, Wang et al. 2001). As a result, regulation of the shunt fraction, along with increases in ventilation and cardiac output, orchestrate the increase in O_2 delivery to the tissues required to support the increase in metabolism.

However, changes in ventilation during the postprandial period are complex. They are regulated by more than just the need to increase O_2 supply. When food is swallowed and reaches the stomach, there is a strong stimulation of the parietal cells to secrete large quantities of HCl for the gastric lumen (Busk et al. 2000a, Wang et al. 2001, Andrade et al. 2004a). Consequently, there is a significant reduction of chloride ions (Cl^-) in the plasma, which leads to a sharp increase in the difference between the concentrations of strong ions (i.e., the millimolar concentration differences between cations and anions), resulting in an increased plasma concentration of bicarbonate ions (HCO_3^-, see Wang et al. 2001, Andrade et al. 2004a). The net result is a wave of alkalization of the blood known as the "alkaline tide." In many amphibians and reptiles, this can last from a few hours to days (Overgaard et al. 1999, Busk et al. 2000a,b, Andersen and Wang 2003, Andrade et al. 2004a, Arvedsen et al. 2005). It has also been observed in many amphibians and reptiles that in parallel with the elevated levels of arterial HCO_3^- during the postprandial period is a sharp increase in the partial pressure of CO_2 ($PaCO_2$). This is the result of CO_2 retention caused by a "relative hypoventilation" (Hicks et al. 2000, Hartzler et al. 2006). Thus, while ventilation increases to enhance O_2 delivery, as has been observed in lizards *Varanus exanthematicus* (Hartzler et al. 2006) and in snakes *P. molurus* (Overgaard et al. 1999), it does not match the increase in metabolism and CO_2 production by the tissues (see also Secor et al. 2000, Skoovgard and Wang 2004). The increasing $PaCO_2$ due to the relative hypoventilation serves to partially compensate the rise in pH as a result of the acid secretion by the stomach epithelium (see Wang et al. 2001). The degree of compensation is inversely proportional to the size of the meal (see Overgaard et al. 1999, Busk et al. 2000a,b, Andrade et al. 2004a, Hartzler et al. 2006). Thus, ingestion of large amounts of food leads to a metabolic alkalosis that is partly compensated by a respiratory acidosis. This respiratory acidosis also minimizes the effects of the metabolic alkalosis on the hemoglobin–oxygen affinity that, due to the Bohr effect, would restrict O_2 unloading at the tissues (Busk et al. 2000a).

Note that other processes are also important in the regulation of HCO_3^- levels in plasma (e.g., the long-term excretion by the kidneys; see Jackson 1986) and that other factors also modulate the levels of ventilation. For instance, Secor et al. (2000) found that pythons reduced their ventilation during the first 24 h after meal ingestion potentially due to alterations in pulmonary mechanics due to the severe restrictions imposed by the size of the prey inside the stomach (Milsom 1988). Studying bullfrogs *Lithobates catesbeianus*, we found that the breathing frequency and the tidal volume were similar between unfed animals and those fed with a meal equivalent to 10% of their body weight (Figure 5.1, unpublished results). However, when exposed to 1% of CO_2 in the inhaled air, the breathing frequency increased in unfed and fed animals, while the tidal volume increased only in unfed animals. Apparently, the large amount of food inside the stomach may restrict the expansion of the lungs, and then only the rise in breathing frequency can contribute to the increase in total ventilation after meal ingestion in bullfrogs.

The sum of the data indicates that control of breathing during this period is very complex. The need for increased O_2 delivery stimulates ventilation while metabolic alkalosis inhibits it. Thus, while total ventilation increases, the air convection requirement (the ratio of total ventilation to oxygen consumption) falls. This leads to the rise in $PaCO_2$ that under

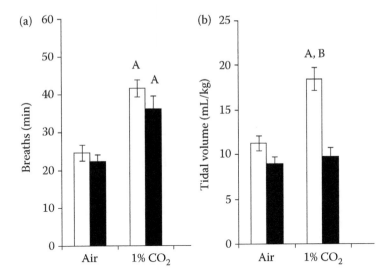

Figure 5.1 Breathing frequency (a) and tidal volume (b) of unfed (white column) and fed (black column) bullfrogs *L. catesbeianus* when exposed to "air" and a mixture of air with "1% of CO_2." Fed animals were measured at 48 h after meal ingestion. The values are mean ± SEM ($N = 17$). (A) Denotes a significant difference of animals breathing air, while (B) denotes a significant difference of unfed animals ($P < 0.05$). (From Carvalho, J. E., D. V. Andrade, and W. K. Milsom, Unpublished results.)

other circumstances would itself be a powerful drive to breathe (Hitzig and Jackson 1978, Branco and Wood 1993, Milsom 1995). The net result depends on a complex interaction between signals arising from the different chemoreceptors responsible for the control of breathing (Milsom 1995, Andrade et al. 2004a). Reptiles possess a wide variety of CO_2/pH-sensitive chemoreceptors including vomeronasal chemoreceptors, intrapulmonary CO_2 chemoreceptors, arterial chemoreceptors (on the pulmonary and carotid arteries as well as on the aorta), and central CO_2/pH-sensitive chemoreceptors (see Milsom 1995, Milsom et al. 2004, Reyes et al. 2014a,b, Reichert et al. 2015). The roles played by these various receptors, as well as the manner in which their inputs are integrated centrally during SDA and the alkaline tide are unknown (Andrade et al. 2004a, Arvedsen et al. 2005). It is not clear what signal(s) drive(s) the initial increase in breathing associated with the increased metabolism. It is also not clear what produces the relative hypoventilation. It is possible that in animals that compensate the effect of metabolic alkalosis during the postprandial period (i.e., the "alkaline tide") with a respiratory acidosis (the "relative hypoventilation"), there is a reduction in sensitivity to the input from the peripheral arterial chemoreceptors (primarily sensing CO_2 levels), an increase in the sensitivity of central pH-sensitive chemoreceptors, or both. In the work that we have conducted with bullfrogs, presented in Figure 5.1, the breathing frequency of unfed and fed animals equally responds to CO_2 in the inhaled air. However, tidal volume did not increase in fed animals, causing a relative hypoventilation under such stimulation with CO_2. We still do not know how (or if) the bullfrogs modulate the chemosensitivity to CO_2 during the postprandial period. In the literature, there are several examples on how animals respond to the changes in CO_2 levels, including amphibians and reptiles (see some examples in Coates and Ballan 1990, Kinkead and Milsom 1996, Coates 2001, Klein et al. 2002, Andrade et al. 2004b), but virtually, nothing has been done with respect to changes in central and peripheral chemosensitivity during the transition to states of elevated metabolism, as is the case for postprandial period. This is an area in need of further investigation.

Concluding remarks

Intermittent feeders provide an intriguing model for investigations of the links between metabolic and cardiorespiratory control. The consequences of feeding on metabolism and respiration depend on the quantity and quality of ingested food (Wang et al. 2005, Greene et al. 2013, Henriksen et al. 2015). In species of amphibians and reptiles that are infrequent feeders, very large meals are accompanied by remarkable changes in the capacity for nutrient absorption in enteric epithelial membranes (Secor 2001, 2009, Secor and Faulkner 2002). The morphological and physiological changes required to deal with large meals greatly increase the metabolic rate and the secretion of gastric acid leading to a systemic arterial alkalosis (Busk et al. 2000a, Wang and Overgaard 2001, Andersen and Wang 2003, Andersen et al. 2003, Wang et al. 2005). This increased need for O_2 supply is somehow sensed and leads to increases in ventilation and cardiac output. The arterial alkalosis is also sensed, putting a brake on the increase in ventilation such that not all the metabolic CO_2 produced is eliminated. This relative respiratory acidosis given by a relative hypoventilation to CO_2 buffers the alkalosis. This relative hypoventilation fails to drive ventilation higher for reasons that are also yet unknown. Research is needed to explore the links between metabolism and cardiorespiratory control that these observations provide.

Acknowledgments

J.E. Carvalho was supported by The São Paulo Research Foundation (FAPESP: 2004/05469-8 and 06/60750-0) and by the National Institute of Research and Technology in Comparative Physiology (INCT Fisio Comp FAPESP: 2008/57712-4). D.V. Andrade was supported by the

National Council for Scientific and Technological Development (CNPq: 302045/2012-0) and by The São Paulo Research Foundation (FAPESP: 2013/04190-9).

References

Andersen, J. B. and T. Wang. 2003. Cardiorespiratory effects of forced activity and digestion in toads. *Physiological and Biochemical Zoology* 76: 459–470.

Andersen, J. B., D. V. Andrade, and T. Wang. 2003. Effects of inhibition gastric acid secretion on arterial acid–base status during digestion in the toad *Bufo marinus*. *Comparative Biochemistry and Physiology* 135A: 425–433.

Andersen, J. B., C. B. Rourke, V. J. Caiozzo, A. F. Bennett, and J. W. Hicks. 2005. Postprandial cardiac hypertrophy in pythons. *Nature* 434: 37–38.

Andrade, D. V., A. P. Cruz-Neto, and A. S. Abe. 1997. Meal size and specific dynamic action in the rattlesnake *Crotalus durissus* (Serpentes: Viperidae). *Herpetologica* 53(4): 485–493.

Andrade, D. V., L. F. Toledo, A. S. Abe, and T. Wang. 2004a. Ventilatory compensation of the alkaline tide during digestion in the snake *Boa constrictor*. *Journal of Experimental Biology* 207: 1379–1385.

Andrade, D. V., G. Tattersall, S. P. Brito, R. Soncini, L. G. Branco, M. L. Glass, A. S. Abe, and W. K. Milsom. 2004b. The ventilatory response to environmental hypercarbia in the South American rattlesnake, *Crotalus durissus*. *Journal of Comparative Physiology* 174B: 281–291.

Arvedsen, S. K., J. B. Andersen, M. Zaar, D. V. Andrade, A. S. Abe, and T. Wang. 2005. Arterial acid–base status during digestion and following vascular infusion of $NaHCO_3$ and HCl in the South American rattlesnake, *Crotalus durissus*. *Comparative Biochemistry and Physiology* 142A: 495–502.

Beaupre, S. J. and D. J. Duvall. 1998. Integrative biology of rattlesnakes. *Bioscience* 48(7): 531–538.

Bennett, A. F. 1994. Exercise performance of reptiles. *Advances in Veterinary Science and Comparative Medicine* 38B: 113–138.

Bennett, A. F. and J. W. Hicks. 2001. Postprandial exercise: Prioritization or additivity of the metabolic responses? *Journal of Experimental Biology* 204: 2127–2132.

Bovo, R. P., A. Fuga, M. A. Micheli-Campbell, J. E. Carvalho, and D. V. Andrade. 2015. Blood oxygen affinity increases during digestion in the South American rattlesnake, *Crotalus durissus terrificus*. *Comparative Biochemistry and Physiology A* 186: 75–82.

Branco, L. G. S. and S. C. Wood. 1993. Effects of temperature on central chemical control of ventilation in the alligator *Alligator mississippiensis*. *Journal of Experimental Biology* 179: 261–272.

Brooks, G. A. 1998. Mammalian fuel utilization during sustained exercise. *Comparative Biochemistry and Physiology* 120B: 89–107.

Brooks, G. A., H. Dubouchaud, M. Brown, J. P. Sicurello, and C. E. Butz. 1999. Role of mitochondrial lactate dehydrogenase and lactate oxidation in the intracellular lactate shuttle. *Proceedings of the National Academy of Sciences of the United States of America* 96: 1129–1134.

Busk, M., F. B. Jensen, and T. Wang. 2000a. Effects of feeding on metabolism, gas transport, and acid–base balance in the bullfrog *Rana catesbeiana*. *American Journal of Physiology* 278: R185–R195.

Busk, M., J. Overgaard, J. M. Hicks, A. F. Bennett, and T. Wang. 2000b. Effects of feeding on arterial blood gases in the American alligator *Alligator mississippiensis*. *Journal of Experimental Biology* 203: 3117–3124.

Clark, T. D., P. J. Butler, and P. B. Frappell. 2005. Digestive state influences the heart rate hysteresis and rates of heat exchange in the varanid lizard *Varanus rosenbergi*. *Journal of Experimental Biology* 208: 2269–2276.

Coates, E. L. 2001. Olfactory CO_2 chemoreceptors. *Respiration Physiology* 129: 219–229.

Coates, E. L. and G. O. Ballan. 1990. Olfactory receptor response to CO_2 in bullfrogs. *American Journal of Physiology* 258(27): R1207–R1212.

Conley, K. E., M. J. Kushmerick, and S. A. Jubrias. 1998. Glycolysis is independent of oxygenation state in stimulated human skeletal muscle *in vivo*. *Journal of Physiology* 511: 935–945.

Crocker-Buta, S. P. and S. M. Secor. 2014. Determinants and repeatability of the specific dynamic response of the corn snake, *Pantherophis guttatus*. *Comparative Biochemistry and Physiology A* 169: 60–69.

Garland, T. and S. J. Arnold. 1983. Effects of full stomach on locomotory performance of juvenile garter snakes (*Thamnophis elegans*). *Copeia* 1983: 1092–1096.

Gladden, L. B. 2001. Lactic acid: New roles in a new millennium. *Proceedings of the National Academy of Sciences of the United States of America* 98: 395–397.

Greene, H. W. 1997. *Snakes: The Evolution of Mystery in Nature*. Berkeley: California Press.

Greene, S., S. McConnachie, S. Secor, and M. Perrin. 2013. The effects of body temperature and mass on the postprandial metabolic responses of the African egg-eating snakes *Dasypeltis scabra* and *Dasypeltis inornata*. *Comparative Biochemistry and Physiology A* 165(2): 97–105.

Hartzler, L. K., S. L. Munns, A. F. Bennett, and J. W. Hicks. 2006. Metabolic and blood gas dependence on digestive state in the Savannah monitor lizard *Varanus exanthematicus*: An assessment of the alkaline tide. *Journal of Experimental Biology* 209: 1052–1057.

Henriksen, P. S., S. Enok, J. Overgaard, and T. Wang. 2015. Food composition influences metabolism, heart rate and organ growth during digestion in *Python regius*. *Comparative Biochemistry and Physiology A* 183: 36–44.

Hicks, J. W. and A. F. Bennett. 2004. Eat and run: Prioritization of oxygen delivery during elevated metabolic states. *Respiration Physiology and Neurobiology* 144: 215–224.

Hicks, J. W., T. Wang, and A. F. Bennett. 2000. Patterns of cardiovascular and ventilatory response to elevated metabolic states in the lizard *Varanus exanthematicus*. *Journal of Experimental Biology* 203: 2437–2445.

Hitzig, B. M. and D. C. Jackson. 1978. Central chemical control of ventilation in the unanesthetized turtle. *American Journal of Physiology* 235: R257–R264.

Hochachka, P. W. and G. N. Somero. 2002. *Biochemical Adaptation: Mechanism and Process in Physiological Evolution*. New York: Oxford University Press.

Holmberg, A., J. Kaim, A. Persson, J. Jensen, T. Wang, and S. Holmgren. 2003. Effects of digestive status on the reptilian gut. *Comparative Biochemistry and Physiology* 133A: 499–518.

Iglesias, S., M. B. Thompson, and F. Seebacher. 2003. Energetics cost of a meal in a frequent feeding lizard. *Comparative Biochemistry and Physiology* 135A: 377–382.

Jackson, D. C. 1986. Acid–base regulation in reptiles. In Heisler, N. (ed.), *Acid–Base Regulation in Animals*, 235–263. Amsterdam: Elsevier Science Publishers.

Jensen, B., C. K. Larsen, J. M. Nielsen, L. S. Simonsen, and T. Wang. 2011. Change of cardiac function, but not form, in postprandial pythons. *Comparative Biochemistry and Physiology A* 160: 35–42.

Jordan, A. D. and J. F. Steffensen. 2005. Specific dynamic action in *Gadus morhua* under normoxia and moderate strong hypoxia. *Comparative Biochemistry and Physiology A* 141: S213.

Kemper, W. F., S. L. Lindstedt, L. K. Hartzler, J. W. Hicks, and K. E. Conley. 2001. Shaking up glycolysis: Sustained, high lactate flux during aerobic rattling. *Proceedings of the National Academy of Sciences of the United States of America* 98: 723–728.

Kinkead, R. and W. K. Milsom. 1996. CO_2-sensitive olfactory and pulmonary receptor modulation of episodic breathing in bullfrogs. *American Journal of Physiology* 270: R134–R144.

Kleiber, M. 1961. *The Fire of Life: An Introduction to Animal Energetics*. New York: John Wiley & Sons.

Klein, W., D. V. Andrade, T. Wang, and E. W. Taylor. 2002. Effects of temperature and hypercapnia on ventilation and breathing pattern in the lizard *Uromastyx aegyptius microlepis*. *Comparative Biochemistry and Physiology* 132A: 847–859.

Lignot, J. H., C. Helmstetter, and S. M. Secor. 2005. Postprandial morphological response of the intestinal epithelium of the Burmese python (*Python molurus*). *Comparative Biochemistry and Physiology A* 141: 280–291.

Luo, Y. and X. Xie. 2008. Effects of temperature on the specific dynamic action of the southern catfish, *Silurus meridionalis*. *Comparative Biochemistry and Physiology A* 149: 150–156.

McCue, M. D. 2006. Specific dynamic action: A century of investigation. *Comparative Biochemistry and Physiology* 144A: 381–394.

McCue, M. D., R. M. Guzman, and C. A. Passement. 2015. Digesting pythons quickly oxidize the proteins in their meals and save the lipids for later. *Journal of Experimental Biology* 218: 2089–2096.

Milsom, W. K. 1988. Control of arrhythmic breathing in aerial breathers. *Canadian Journal of Zoology* 66: 99–108.

Milsom, W. K. 1995. The role of CO_2/pH chemoreceptors in ventilatory control. *Brazilian Journal of Medical and Biological Research* 28: 1147–1160.

Milsom, W. K., A. S. Abe, D. V. Andrade, and G. J. Tattersall. 2004. Evolutionary trends in airway CO_2/H^+ chemoreception. *Respiration Physiology and Neurobiology* 144: 191–202.

Navas, C. A., F. R. Gomes, and J. E. Carvalho. 2008. Thermal relationships and exercise physiology in anuran amphibians: Integration and evolutionary implications. *Comparative Biochemistry and Physiology* 151: 344–362.

Overgaard, J., J. B. Andersen, and T. Wang. 2002. The effects of fasting duration on the metabolic response to feeding in *Python molurus*: An evaluation of the energetic costs associated with gastrointestinal growth and upregulation. *Physiological and Biochemical Zoology* 75(4): 360–368.

Overgaard, J., M. Busk, J. W. Hicks, F. B. Jensen, and T. Wang. 1999. Respiratory consequences of feeding in the snake *Python molurus*. *Comparative Biochemistry and Physiology* 124: 359–365.

Owen, S. F. 2001. Meeting energy budgets by modulation of behavior and physiology in the eel (*Anguilla anguilla* L.). *Comparative Biochemistry and Physiology* 128A: 631–644.

Preest, M. R. 1994. Sexual size dimorphism and feeding energetics in *Anolis caroli-nensis*: Why do females take smaller prey than males? *Journal of Herpetology* 28(2): 292–298.

Reichert, M., D. L. Brink, and W. K. Milsom. 2015. Evidence for a carotid body homolog in the lizard *Tupinambis merianae*. *Journal of Experimental Biology* 218: 228–237.

Reyes, C., A. Y. Fong, D. L. Brink, and W. K. Milsom. 2014a. Distribution and innervation of putative arterial chemoreceptors in the bullfrog (*Rana cates-beiana*). *Journal of Comparative Neurology* 522: 3754–3774.

Reyes, C., A. Y. Fong, D. L. Brink, and W. K. Milsom. 2014b. Distribution and innervation of peripheral arterial chemoreceptors in the red eared slider (*Trachemys scripta elegans*). *Journal of Comparative Neurology* 522: 3753–3774.

Secor, S. M. 2001. Regulation of digestive performance: A proposed adaptive response. *Comparative Biochemistry and Physiology* 128A: 565–577.

Secor, S. M. 2003. Gastric function and its contribution to the postprandial meta-bolic response of the Burmese python *Python molurus*. *Journal of Experimental Biology* 206: 1621–1630.

Secor, S. M. 2009. Specific dynamic action: A review of the postprandial metabolic response. *Journal of Comparative Physiology B* 179: 1–56.

Secor, S. M. and J. Diamond. 1995. Adaptive responses to feeding in Burmese pythons: Pay before pumping. *Journal of Experimental Biology* 198: 1313–1325.

Secor, S. M. and J. Diamond. 1997. Effects of meal size on postprandial responses in juvenile Burmese pythons (*Python molurus*). *American Journal of Physiology* 272: R902–R912.

Secor, S. M. and J. Diamond. 1998. A vertebrate model of extreme physiological regulation. *Nature* 395: 659–662.

Secor, S. M. and M. Boehm. 2006. Specific dynamic action of ambystomatid sal-amanders and the effects of meal size, meal type, and body temperature. *Physiological and Biochemical Zoology* 79: 720–735.

Secor, S. M. and J. Diamond. 2000. Evolution of regulatory responses to feeding in snakes. *Physiological and Biochemical Zoology* 73: 123–141.

Secor, S. M. and A. C. Faulkner. 2002. Effects of meal size, meal type, body tem-perature, and body size on the specific dynamic action of the marine toad, *Bufo marinus*. *Physiological and Biochemical Zoology* 75(6): 557–571.

Secor, S. M., D. Fehsenfeld, J. Diamond, and T. E. Adrian. 2001. Responses of python gastrointestinal regulatory peptides to feeding. *Proceedings of the National Academy of Sciences of the United States of America* 98(24): 13637–13642.

Secor, S. M., J. W. Hicks, and J. Diamond. 2000. Ventilatory and cardiovascular responses of a python (*Python molurus*) to exercise and digestion. *Journal of Experimental Biology* 203: 2447–2454.

Secor, S. M., E. D. Stein, and J. Diamond. 1994. Rapid upregulation of snake intes-tine in response to feeding: A new model of intestinal adaptation. *American Journal of Physiology* 266: G695–G705.

Secor, S. M., J. A. Wooten, and C. L. Cox. 2007. Effects of meal size, meal type, and body temperature on the specific dynamic action of anurans. *Journal of Comparative Physiology B* 177: 165–182.

Seeherman, H. J., C. R. Taylor, G. M. Maloiy, and R. B. Armstrong. 1981. Design of the mammalian respiratory system. II: Measuring maximum aerobic capac-ity. *Respiration Physiology* 44(1): 11–23.

Shulman, R. G. and D. L. Rothman. 2001. The "glycogen shunt" in exercising muscle: A role for glycogen in muscle energetics and fatigue. *Proceedings of the
National Academy of Sciences of the United States of America* 98: 457–461.

Skoovgard, N. and T. Wang. 2004. Cost of ventilation and effect of digestive state
on the ventilatory response of the tegu lizard. *Respiration Physiology and
Neurobiology* 141: 85–97.

Starck, J. M. and K. Beese. 2001. Structural flexibility of the intestine of Burmese
python in response to feeding. *Journal of Experimental Biology* 204: 325–335.

Starck, J. M. and K. Beese. 2002. Structural flexibility of the small intestine and liver
of garter snakes in response to feeding and fasting. *Journal of Experimental
Biology* 205: 1377–1388.

Suarez, R. K. 1996. Upper-limits to mass-specific metabolic rates. *Annual Review of
Physiology* 58: 583–605.

Suarez, R. K. 1998. Oxygen and the upper limits to animal design and performance. *Journal of Experimental Biology* 201: 1065–1072.

Taigen, T. L. and K. D. Wells. 1985. Energetics of vocalization by an anuran
amphibian (*Hyla versicolor*). *Journal of Comparative Physiology* 155B: 163–170.

Toledo, L. F., A. S. Abe, and D. V. Andrade. 2003. Temperature and meal size effects
on the postprandial metabolism and energetics in a boid snake. *Physiological
and Biochemical Zoology* 76(2): 240–246.

Waas, S., R. A. Werner, and J. M. Starck. 2010. Fuel switching and energy partitioning during the postprandial metabolic response in the ball python (*Python
regius*). *Journal of Experimental Biology* 213: 1266–1271.

Wang, Q., W. Wang, Q. Huang, Y. Zhang, and Y. Luo. 2012. Effect of meal size
on the specific dynamic action of the juvenile snakehead (*Channa argus*).
Comparative Biochemistry and Physiology A 161: 401–405.

Wang, T. 2001. Physiological consequences of feeding in animals. *Comparative
Biochemistry and Physiology* 128A: 395–396.

Wang, T. and J. W. Hicks. 1996. The interaction of pulmonary ventilation and the
right–left shunt on arterial oxygen levels. *Journal of Experimental Biology* 199:
2121–2129.

Wang, T., M. Busk, and J. Overgaard. 2001. The respiratory consequences of feeding in amphibians and reptiles. *Comparative Biochemistry and Physiology*
128A: 535–549.

Wang, T., W. W. Burggren, and E. Nobrega. 1995. Metabolic, ventilatory, and acid–
base responses associated with specific dynamic action in the toad *Bufo
marinus*. *Physiological Zoology* 68: 192–205.

Wang, T., J. B. Andersen, and W. Hicks. 2005. Effects of digestion on the respira-
tory and cardiovascular physiology of amphibians and reptiles. In Starck,
J. M. and T. Wang (eds.), *Physiological and Ecological Adaptations to Feeding in
Vertebrates*, pp. 279–303. Enfield, New Hampshire: Science Publishers.

Weibel, E. R., L. D. Bacigalupe, B. Schmitt, and H. Hoppeler. 2004. Allometric
scaling of maximal metabolic rate in mammals: Muscle aerobic capacity as
determinant factor. *Respiration Physiology and Neurobiology* 140: 115–132.

Wood, S. C. 1984. Cardiovascular shunts and oxygen transport in lower vertebrates. *American Journal of Physiology* 247: R3–R14.

chapter six

Temperature effects on the metabolism of amphibians and reptiles
Caveats and recommendations

Denis Vieira de Andrade

Contents

Introduction

All animals need energy. Indeed, each and every activity, or behavior, an animal will engage in throughout its lifetime will require the use of energy. As in the majority of other animals, Squamate reptiles derive their energy from the breakdown of organic substrates, often referred to as catabolism, acquired through feeding. Adenosine triphosphate (ATP) will eventually be hydrolyzed to adenosine diphosphate (ADP) and P_i, releasing the energy stored in its chemical bonds to power subcellular structures, cells, tissues, organs, and systems, which, acting under neural and/or humoral control, will be manifested as identifiable activities and behaviors (see Chapter 1). Thus, by focusing on metabolism, one is in a privileged position to appreciate the interactions of a given organism in terms of energy exchange with the biotic and abiotic components of its particular environment. This is of indisputable value as it can reveal current and future tradeoffs in terms of functional and energetic constraints with important ecological and evolutionary consequences (see Bennett 1982; Congdon et al. 1982; Nagy 1983; Pough et al. 1992; Nagy et al. 1999; McNab 2002; Suarez 2012). Putting it simply, every living organism requires energy to exist and the patterns of its acquisition and utilization are fundamental for the success of any given species.

It follows that the relevance of metabolic measurements has always been acknowledged by animal physiologists, including those interested in amphibians and reptiles. As a consequence, metabolic rate determination is, quite probably, one of the most intense and widespread physiological parameters quantified for any animal group (Benedict 1932; Kleiber 1947, 1961; Bennett and Dawson 1976; Schmidt-Nielsen 1984; MacNab 2002; Suarez 2012). Furthermore, aerobic metabolism can be determined based on the rate of oxygen consumption, which is relatively uncomplicated, accurate, and not an overly expensive methodology (Lighton 2008). Accordingly, in this chapter, I will focus entirely on aerobic metabolism and will use rates of oxygen uptake and metabolism interchangeably, as this will suffice to tackle the main goals of this review. This does not imply, to any extent, the denial of the relevance of anaerobic metabolism for the behavioral ecology and physiology of amphibians and reptiles, and readers are advised to consider some early, but now classic, studies focusing on this particular aspect (Bennett 1972; Bennett and Licht 1974; Ruben 1976; Feder and Arnold 1982; Gleeson 1991).

Metabolic rate is influenced by a number of intrinsic features dependent on the animal being measured (e.g., body mass, sex, and fed state), as well as by external parameters, usually biotic or abiotic variables from the particular habitat where the animal is found (see McNab 2002). Among the physical abiotic variables known to exert an influence on the metabolic rate of animals, temperature can easily be identified as the one whose effects have been largely investigated. This is justifiable, as temperature is widely recognized as the single physical parameter with the most profound impacts on animal function (Huey 1982; Angilletta 2009; Tattersall et al. 2012). In the case of ectothermic animals, including amphibians and reptiles, the influence of temperature on metabolism might be especially important in an ecological and evolutionary context when compared to endothermic animals (Angilletta et al. 2002). The metabolic rate of amphibians and reptiles is orders of magnitude lower than similar sized endothermic vertebrates (Bennett 1978, 1980; Else and Hulbert 1981; Pough 1983; di Prampero 1985; Else and Hulbert 1985; Bennett and Harvey 1987; Bennett 1994; Hedrick et al. 2015) meaning that, along with the lack of effective insulation, their capacity to use metabolically derived heat for body temperature regulation is usually (but not always, as discussed later in this chapter) negligible (see Seebacher et al. 2005). As a consequence, body temperature regulation in amphibians and reptiles is highly dependent on external heat sources and the behavioral exploration of the different thermal niches available in a particular habitat (Bartholomew 1982; Huey 1982; Hutchison and Dupré 1992; Seebacher et al. 2005). This thermoregulatory strategy is often conducive to a daily or seasonal thermal cycle in which animals naturally are exposed to very large fluctuations in body temperature and changes in activity level (Huey and Pianka 1977; Avery 1982; Gregory 1982; Pinder et al. 1992; Abe 1995; Carvalho et al. 2010; Gunderson and Leal 2015; Sanders et al. 2015). For example, differences of more than 10°C in the body temperature during a diurnal cycle are quite commonly experienced by many species of amphibians and reptiles (Brattstrom 1963, 1965; Hutchison and Dupré 1992; Andrade et al. 2004). All these factors taken into consideration underscore the relevance of quantifying the effects of temperature on the metabolic rate of these particular animals.

Indeed, temperature effects on the metabolic rate have been thoroughly quantified in diverse representatives of amphibians and reptiles from the early days of comparative animal physiology up to the present (Benedict 1932; Bennett and Dawson 1976; Andrews and Pough 1985; Gatten et al. 1992; Rome et al. 1992; Hillman et al. 2009). As is true for other ectothermic organisms, the metabolism of amphibians and reptiles usually increases with body temperature for the temperature range within their thermal tolerance limits (Bennett and Dawson 1976; Gatten et al. 1992; Halsey et al. 2015). Beyond these limits, functional erosion will happen and disrupt the relationship between physiological function and temperature (Huey and Stevenson 1979). Within the limits of thermal tolerance, the metabolism of amphibians and reptiles will increase with temperature, often linearly, approximately doubling for each 10°C increase in body temperature; that is, the Q_{10} value usually hovers around 2 (Bennett and Dawson 1976; Gatten et al. 1992; White et al. 2006; Halsey et al. 2015). As body temperature regulation in amphibians and reptiles is dissociated from metabolic heat production, the temperature effect on their metabolism is assumed to be a mostly passive thermodynamic consequence of the temperature on the biochemical reactions sustaining the different physiological systems (see Gillooly et al. 2001), including those sustaining changes in activity levels (see Halsey et al. 2015). The sum of the energetic expenditures of all these reactions is nothing more than the metabolic rate itself. Although the same can be said for any living organism, there is a fundamental difference between ectothermic and endothermic organisms on how metabolism varies as a function of temperature (Bennett 1980; Bartholomew 1982; Pough 1983). This is related primarily to the fact that endotherms usually (but not always) will defend a constant body temperature with the expense of metabolic expenditure, a case usually referred to as homeothermic endothermy (Lowell and Spiegelman 2000; McNab 2002). In such cases, there will be a lower and upper critical temperature below and above which, respectively, animals will spend a surplus energy to keep their body temperature constant. In the middle temperature interval set by the lower and upper critical temperatures, animals can modulate the heat being produced with the amount of heat being exchanged with the environment, keeping their body temperature constant with little, or negligible, change in metabolism. This is referred to as the thermoneutral zone (Nichelmann and Tzschentke 1995).

While the concept of thermoneutral zone has long been incorporated in the measurements of the temperature effects on the metabolism of endothermic organisms, the peculiarities of body temperature regulation of amphibians and reptiles has often been neglected by those measuring their metabolic rate. The current paradigm for the assessment of the effects of temperature variation on the metabolism of amphibians and reptiles is the measurement of animals submitted to constant temperature regimes for a pre- or postestablished experimental period that can extend for hours or even days. In some cases, the duration of the experiment and even the choice of the experimental temperatures are often determined without a proper consideration of the thermal biology of the animals being measured. As a consequence, animals that normally experience considerable variation in body temperature during the diurnal cycle for their entire life are submitted, by default, to a constant temperature regime while having their metabolic rate determined. As experiments may last for periods up to many days, animals are then subjected to experimental conditions that they most likely never experience in nature. Although previous data on the metabolism of amphibians and reptiles obtained under constant temperature regimes remain relevant (see Benedict 1932; Bennett and Dawson 1976; Andrews and Pough 1985; Gatten et al. 1992; Rome et al. 1992; Hillman et al. 2009; and references therein), more solid and reliable insights about energy use may be attained by narrowing the gap between methodological protocols and the thermal biology of the experimental organisms (see Newman et al. 2015 for a similar approach on another subject). Thus, goals of this chapter can be defined as (1) a plea for the incorporation of thermal biology information in the assessment of the temperature effects on the metabolism of amphibians and reptiles; (2) an alert that temperature effects on metabolism can be associated with subtle attributes of temperature variation rather than to plain differences in mean averages; and (3) a discussion on the methodological caveats, and potential ways to get around them, involved in the examination of the temperature effects on the metabolism of amphibians and reptiles.

Body temperature variation

As true for any ectothermic organism, the body temperature of amphibians and reptiles is highly dependent on the prevalent ambient thermal conditions, which are explored predominantly via behavioral adjustments

(Huey and Pianka 1977; Seebacher et al. 2005; Pough et al. 2015). Thermoregulatory costs and benefits are integrated for a given set of conditions and tradeoffs resulting in different thermoregulatory patterns (Huey and Slatkin 1976) and activity (Gunderson and Leal 2015). Obviously, this response also varies considerably according to thermal characteristics of the environment and with the thermoregulatory capabilities of the animal being considered. For example, the high specific heat capacity of water makes it almost impossible for the majority of the aquatic amphibians or reptiles to regulate their body temperature at a temperature different from that of the water body where they happen to be immersed (Hillman et al. 2009). For the body size range typical of most amphibians and reptiles, any thermal difference will rapidly equilibrate with the water as they lack any effective insulation. We can also attribute diel and seasonal changes in body temperature that are minimized for those organisms thriving in the aquatic environments, compared to those found in terrestrial habitats, to the physical properties of the water (Tracy 1976, 1982; Spotila et al. 1992). There are, however, few remarkable exceptions, even in aquatic habitats. Leatherback sea turtles and large crocodiles, for example, are able to use their large body size, coupled with metabolic and cardiovascular adjustments, to keep their body temperature relatively constant despite considerable changes in the thermal environment (Fray et al. 1972; Seebacher et al. 1999; Bostrom et al. 2010). Leatherbacks are, indeed, able to keep their body temperature well elevated above sea water temperature (Paladino et al. 1990; Casey et al. 2014). Finally, incorporating historical and adaptive factors, different species will be more dedicated to buffering body temperature from change in the thermal environment (commonly referred to as active thermoregulators), while others will be more relaxed in terms of allowing greater body temperature variations (referred to as thermoconformers) (Huey and Slatkin 1976).

In general, amphibians, with their predominantly nocturnal activity, their close association with water, and their moist skin, which constrain some thermoregulatory possibilities, are biased to the thermoconformity end of the scale (Tracy 1975; Hillman et al. 2009). Exceptions exist (Shoemaker et al. 1987, 1989; Tattersall et al. 2006), however, including some strategies for regulating the temperature of the environment of developing eggs (Méndez-Narváez et al. 2015). Reptiles, on the other hand, present a more even distribution on the thermoregulatory spectrum, with diurnal heliothermic lizards representing the epitome of active thermoregulation (Huey 1982). Abandoning our mammalian paradigm, an important point to be made is that an accurate thermoregulation for an ectotherm organism does not necessarily mean a higher and constant body temperature. While a higher constant body temperature does indeed favor the performance of many activities that depend on muscle contraction, it is equally true that lower body temperatures favor important aspects of life history. Most importantly for the sake of this chapter, lower body temperatures correspond with lower metabolic rate and an associated savings in energy expenditure.

In conclusion, a fundamental step in the assessment of the effects of temperature on the metabolism of amphibians and reptiles is the assessment of variation in body temperature experienced by these animals under natural settings. Ideally, one should accomplish this by incorporating data on activity within a daily and seasonal time scale. Sadly, such studies are exceedingly scarce, especially for amphibians.

Value of incorporating thermal biology into metabolic determinations

Most estivating or hibernating species exhibit significant changes in mean body temperature, and they also exhibit marked changes in their daily pattern of body temperature variation. For example, in the black and white South American tegu lizards, *Salvator merianae*, body temperature exhibits a marked variation along the daily cycle during the hot and rainy season (roughly corresponding to the spring and summer) in Southeastern Brazil (Andrade et al. 2004; Milsom et al. 2012; Sanders et al. 2015). Similar changes in activity and body temperature are widespread in many other species of reptiles and amphibians (Gregory 1982; Abe 1995). In the specific case of *S. merianae*, active lizards emerge from their nightly retreat early in the morning to bask, resulting in a rapid rise in body temperature accompanied by a remarkable increase in heart rate (Andrade et al. 2004; Sanders et al. 2015). Tegus then keep their body temperature high, around 35°C, for the period they are active. Afterwards, in late afternoon, lizards halt activity and return to the shelter where they will spend the night. At this time, a massive and almost instantaneous drop in heart rate is observed, even before any change in body temperature is noticed. As a consequence, much of the heat gained during the day is trapped within the lizard's body and is lost only very slowly throughout the night (Andrade et al. 2004; Sanders et al. 2015), although some thermogenesis may also contribute (discussed in the next section). On the other hand, during the cold and dry winter months, tegus spend the entire season in a dormant state hidden in shelters dug in the soil. Under this condition, body temperature equilibrates with shelter temperature and daily fluctuations in body temperature are compressed to the level dictated by the fluctuations in the shelter temperature, which are minor since shelters are usually well insulated (Milsom et al. 2012).

The detailed description of the tegu's thermal biology just discussed was only obtained years later, after some researchers, myself included, had examined the seasonal variation in a number of physiological parameters, including the seasonal changes in metabolism (Abe 1983, 1995; Andrade and Abe 1999; Andrade et al. 2008a,b). The problem was that at the onset of such investigations, tested temperatures were established not on the basis of the thermal biology of the lizards themselves but on the average temperature of their shelters (i.e., 17°C and 25°C). As a consequence, half of our results were obtained at a temperature (25°C) that coincides only briefly twice a day with the body temperature of active tegus as it varies up and down along its circadian cycle, while the other half, obtained at the lower temperature of 17°C, completely fails to reflect the body temperatures experienced by the active tegus, at any given time (Figure 6.1). For the dormancy period, as the animals allow their body temperature to equilibrate with their retreat shelters, the lower temperature of 17°C agrees well with the temperature experienced by the animals. On the other hand, the higher experimental temperature of 25°C is never experienced by any dormant lizards (see Andrade et al., 2004). Adjusting experimental temperatures across different seasons was a necessary compromise for comparative purposes. However, I regret that some of the previous physiological measurements on this lizard species were not made in temperatures more realistically bounded to its thermal biology. Undoubtedly, the validity of any discussion on aspects of animal temperature and energetics is certainly compromised when metabolic determinations are taken

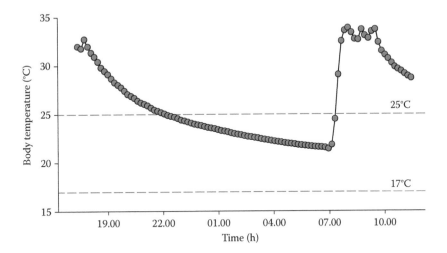

Figure 6.1 Daily variation in the body temperature of an adult tegu lizard (*S. merianae*) (red circles) superimposed on arbitrarily chosen experimental temperatures (blue dashed lines). The graph illustrates how the choice of experimental temperatures without proper consideration of the thermal biology of the animal under study may seriously compromise data interpretation (see text for details). Body temperature data were recorded by radio-telemetry during the season of activity for this species under seminatural conditions (see Sanders et al. 2015).

at temperatures that the animal never experiences or experiences only briefly during the day. Clearly, solid ecophysiological inferences involving the examination of temperature effects on metabolism demand a careful consideration of the thermal biology of the organism being studied (see also Dabruzzi et al. 2012) and such an approach, although seemingly obvious, should be actively encouraged.

Temperature variation and metabolism

If the choice of experimental temperatures may have some obvious consequences for assessing the temperature effects on the metabolism of amphibians and reptiles, differences in the thermal regime, between those normally experienced by the animals and the one to which they are subjected during experimental measurements, may have more subtle and largely neglected effects. The central difference here lies in the fact that while most amphibians and reptiles experience a circadian cycle of body temperature variation (Huey and Pianka 1977; Bartholomew 1982; Huey 1982; Hutchison and Dupré 1992), metabolic measurements have traditionally been made under a constant thermal regime (Bennett and Dawson, 1976; Andrews and Pough 1985; Gatten et al. 1992; Rome et al. 1992). While the potential caveats associated with the neglect of thermal regimes are uncertain, we can expect that they are likely to be more pronounced in those species showing more marked thermal cycles (see Kearn et al. 2015) and for experimental protocols demanding longer measurement durations, such as many days. Constant temperature measurements are certainly adequate to answer specific questions and have contributed to our understanding of the intricate interplay between metabolism and temperature in amphibians and reptiles. However, extrapolations based on metabolic measurements taken under constant temperature into an ecological/behavioral framework, in which animals normally experience a fluctuating thermal regime, might be potentially misleading, particularly when examining processes over longer time scales (see also Kingsolver et al. 2015; Stahlschmidt et al. 2015).

Daily thermal cycles are known to influence a number of physiological functions/rhythms, often in association with light-dark cycles (Underwood 1984). However, surprisingly few studies have being dedicated to the effects of thermal cycles on metabolism. Gavira and Andrade (2013a) examined the effect of a fluctuating thermal regime (12:12 h, 20–30°C) compared to the equivalent constant temperature (i.e., 25°C) during the digestion of a neotropical pitviper, *Bothrops alternatus*, a process that extends for many days (Figure 6.2). Their results showed that both the resting metabolic rate measured before feeding and the maximum rate measured during the meal digestion were lower in those snakes measured under the fluctuating thermal regime. However, as the duration of the digestion was longer

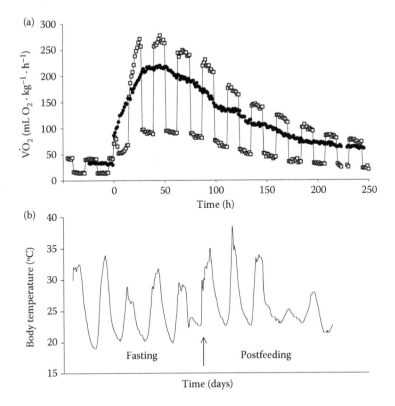

Figure 6.2 (a) The postprandial metabolic response of the neotropical viperid snake *B. alternatus* fed with a mouse meal equaling 30% of the snake's body mass under a constant thermal regime of 25°C (filled circles) and under a 12:12 h fluctuating thermal regime from 20°C to 30°C (open squares). The metabolic response recorded under the latter regime reflects more realistically the body temperature variation normally experienced by snakes (as shown in b). Time "0" indicates the time of feeding; samples at every 70 min; $n = 8$ (see Gavira and Andrade 2013a). (b) Body temperature variation recorded from an adult rattlesnake, *C. durissus*, few days prior and after feeding a rat meal equaling to 30% of its body mass (D.V. Andrade, unpublished data).

in this same group, the final energetic cost of meal digestion did not differ between the different thermal regimes. These results allowed for the discussion of some important ecological consequences of the temperature effects on the digestion of snakes that would not be possible by examination only under constant thermal regimes. Indeed, the initial impetus to perform this study was the observation that captive rattlesnakes (*Crotalus durissus*) kept in outdoor pens were able to finish meal digestion even when experiencing nighttime temperatures as low as 6°C, as long as they could freely thermoregulate in the intervening days, while, under experimental conditions they would usually fail to complete digestion at constant thermal regimes as high as 20°C (Andrade, unpublished observation).

Focusing on the effects of thermal regime on the resting metabolic rate of *B. alternatus* clearly show that animals under a fluctuating regime have a lower metabolism than those submitted to the constant regime. For example, the metabolic rates measured at constant 20°C and 30°C were 28% and 18% greater than the rates measured at these same temperatures when they fluctuate between each other by a 12 h interval (Gavira and Andrade 2013a). In another viperid species, the South American rattlesnake, *C. durissus*, we have found that the lowering of metabolism under fluctuating thermal regimes compared to constant ones becomes more pronounced as the temperature interval considered shifts to higher temperatures (Fabrício-Neto et al., unpublished data). One possible explanation for these observations may involve changes in activity level and, thus, in the accompanying level of metabolism as the temperature increases. In this regard, Halsey et al. (2015) demonstrated, in insects and crustaceans, that the temperature effect on metabolic rate incorporates changes in metabolism that were associated with changes in activity level rather than to temperature *per se*. We do not know if such observations are valid for amphibians and reptiles and whether this may explain differences in metabolism between different thermal regimes. However, it seems plausible to expect that differences in the level and recurrence of activity cycles might occur between animals kept under a constant or fluctuating thermal regime and, therefore, affect metabolic measurements (see Andrade et al. 2008b). Finally, it remains unknown whether submitting the animals to experimental conditions more diverse from their normal thermal regimes causes any stress-associated responses, which, in turn, might influence the level of metabolic and/or locomotory activity.

Body temperature variation encompasses more complex changes than could be assessed by experimentally submitting animals to constant levels of different mean temperatures (Vázquez et al. 2015). Also, responses to complex and more realistic regimes in temperature variation can be modulated by acclimation processes and shifts in thermoregulatory behavior (see Chapter 2; see also Angilletta 2009; Huey et al. 2012; Basson and Clusella-Trullas 2015), with important interactive consequences for growth, metabolism, osmoregulation (Davies et al. 2015; Stahlschmidt et al. 2015), and, potentially, to the evolution of physiological adaptation and conservation (see Sunday et al. 2014; Agustín et al. 2015; Buckley et al. 2015). Thus, thermal regime, temperature variance, distribution of thermal microclimates, occurrence of extreme values, and predictability of temperature changes are all relevant attributes associated with body temperature variations that have been recently acknowledged as equally important, sometimes even more relevant, as differences in average values (see, e.g., Ketola et al. 2012; Caillon et al. 2014; Manenti et al. 2014;

Caldwell et al. 2015; Dowd et al. 2015; Ketola and Saarinen 2015; Ma et al. 2015; Turriago et al. 2015; Vázquez et al. 2015). Indeed, Sears and Angilletta (2015) showed that both the heterogeneity and spatial structure of the temperature distribution in the habitat are important components to understand the potential effects of temperature changes on organismal performance. Also, responses to temperature may differ with life stages, and conditions experienced early in life may result in long-lasting later consequences (Levy et al. 2015; Turriago et al. 2015). For example, Horne et al. (2014) showed that differences in diel temperature variance during the embryonic development of the sea turtle *Caretta caretta* cause significant phenotypic changes, which potentially influence their survival. In the snake *Notechis scutatus*, Aubret and Shine (2009) observed that raising juveniles under different temperature treatments makes them adjust to their respective thermal regimes but, at the same time, decreases their later potential for adjusting to a sudden shift in ambient conditions. Also in snakes, Lorioux et al. (2012) found that suboptimal thermal regimes influenced hatchling traits in the Children's python, *Antaresia childreni*, by maternal effects. Thus, the basic idea brought into consideration in this section is that before examining the temperature effects on any biological parameter, we should first question the attributes of the temperature variation itself and, if possible, determine the thermal history of the study animals. The recognition that such aspects are of particular relevance in the realm of conservation physiology (see Chapter 7; Carey 2005; Wikelski 2006; Niehaus et al. 2012; Kingsolver et al. 2015), especially given changes in the thermal environment, both at global and microhabitats scales, is one of the most recognizable footprints associated with modern human activity (IPCC 2014) with some predicted disastrous consequences for biodiversity (Sinervo et al. 2010). Therefore, the examination of the energetic interrelations of animals (i.e., metabolic determinations) under variable and dynamic thermal conditions and in combination with information on their previous thermal history may provide a better and more realistic means to evaluate the consequences of anthropogenic influences on the environment and how animals might respond to them.

Methodological caveats and recommendations

Although metabolic rates are almost always inferred from the determination of rates of oxygen uptake, I will not discuss specific problems (and possibilities) associated with respirometry. For this purpose, readers are referred to Lighton (2008). Herein, I will focus uniquely on some potential confounding factors intrinsic to the assessment of temperature effects on metabolic determinations. The recognition of such problems is neither new nor original (see Benedict 1932; Bartholomew 1982) and, in many instances, the potential to bias data interpretation is indeed negligible, as has been traditionally assumed (Bartholomew 1982; Fraser and Grigg 1984). However, as physiological data are being incorporated into predictive scenarios involving changes of a few degrees celsius in environmental temperatures, minor uncertainties in the assessment of temperature effects on metabolism may actually become nonnegligible (see similar discussion in Bakken and Angilletta 2014). This becomes especially true as the empirical evaluation of temperature effects on organismal functions, as well as the modeling of it, began to be framed on the basis of scenarios of temperature change of a few degrees over a projected and dilated time scale. Indeed, the assessment of the effects of minor changes in body temperature may prove to be an approach more realistically linked to the predicted alterations in the physical environment (see Kearney and Porter 2004; Kearney et al. 2009). As a consequence, the reliability of such approaches will forcedly depend on the accuracy of adequately controlling and monitoring the body temperature of the animals under examination.

There are few experimental protocols that appear to be as simple and straightforward as the examination of the temperature effects on metabolism. Nonetheless, even fewer protocols are haunted by so many caveats. A most common procedure will be as follows: animals are placed inside respirometry chambers, chambers are placed inside temperature-controlled environments, measurements are taken, temperature is changed, measurements are repeated, and any differences found in the results are assumed to reflect differences occurring between the two temperatures tested. However, controlling the temperature of the climatic chambers and/or rooms does not ensure that the body temperature of the experimental animal is at that same temperature. Many factors can contribute to the occurrence of a differential between body and environmental temperature. Enough time has to be allowed for the animals to equalize their temperature with the environment and this is influenced by the dynamics of temperature change and by body size. Larger animals will require longer periods to reach thermal equilibria; moreover, vasomotor adjustments, such as peripheral vasoconstriction, can extend this period even further (Seebacher et al. 2005). For example, a rattlesnake with a body mass around 1 kg can take 4–6 h to reach thermal equilibrium as ambient temperature is shifted between 20°C and 30°C, and this period varies with the direction of temperature change (Fabrício-Neto et al., unpublished data; Figure 6.3). If ambient temperature is dynamically changed at faster rates (e.g., 12°C/h), this same snake will never reach thermal equalization (Figure 6.3). A 3 kg tegu lizard, *S. merianae,* can withstand significant temperature differentials up to 12 h due to its larger body mass and remarkable vasomotor response (Sanders et al. 2015). Thus, under protocols focusing on rapid temperature changes, assumed body temperature (based on the control of ambient temperature) will be considerably different from the actual body

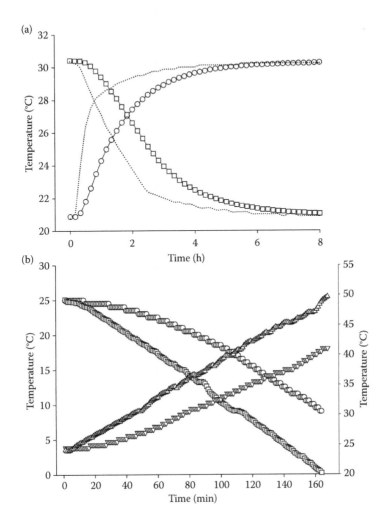

Figure 6.3 (a) Body temperature variation of an adult rattlesnake (*C. durissus*) as environmental temperature (dotted lines) is changed from 20°C to 30°C (open squares) and from 30°C to 20°C (open circles). Notice that body temperature requires many hours to equalize with environmental temperature and that this time is affected by the direction of temperature change. (b) The body temperature of the same snake species (open circles and inverted triangles) never reaches equalization if environmental temperature (open squares and triangles) is changed at a rapid (12°C/h) rate. (a) Depicts mean values of four individual snakes, with dots 10 min apart from each other. (b) Depicts data from one individual snake during warming (open circles and squares) and cooling (open triangles and inverted triangles) trials; dots are 1 min apart; temperature change for both curves start at 25°C, at time 0. All data were recorded from snakes surgically implanted with temperature data loggers. (Courtesy of R.S.B. Gavira and A. Fabrício-Neto, unpublished data.)

temperature experienced by the animals, and this effect is magnified by larger body sizes (see also McNab and Auffenberg 1976; Auffenberg 1981; Tracy 1982; Spotila et al. 1992).

Even when animals are allowed enough time to reach thermal equilibrium with their surroundings, body temperature may not be exactly the same of the environment (see Sunday et al. 2014). Although that might be accurate for most ectothermic organisms, in most instances, particularly for small-bodied reptiles (Fraser and Grigg 1984), there are cases in which temperature differentials will occur. Live animals are continuously engaged in different physiological processes and have thermal properties inherently different from the physical medium surrounding them, which, in turn, affect their rates of heat exchange and ultimately their body temperature (see a similar criticism for determining operative body temperatures in Seebacher and Shine 2004). Amphibians, in general, have a highly permeable skin that makes them particularly susceptible to elevated rates of evaporative water loss (Hillman et al. 2009). As water evaporates, the heat required for the phase change, that is, the heat of vaporization, is taken up from the surface of the material where it happens to occur (the animal skin in this case) causing its temperature to decrease by evaporative cooling. Thus, animals with high rates of evaporative water loss, such as amphibians, will usually exhibit a body surface temperature cooler than the surrounding medium (Brattstrom 1963; Bartelt and Peterson 2005). This effect will be greater at low air humidity and high temperatures, following an increase in the potential for water to vaporize (see Gates 1980). Whether the evaporative cooling happening on the surface of the animal's body will affect core body temperature will depend on other components involved in thermal equilibration (Tracy 1975, 1976; Hillman et al. 2009). In the case of reptiles, with generally low rates of cutaneous evaporative water loss, evaporative cooling will occur mainly from the evaporation from the respiratory surface and will have, supposedly, low potential to influence body temperature (Bartholomew 1982).

A relatively frequent misconception about the attributes of ectothermic organisms is the false belief that they do not produce heat. In other words, only true endotherms are capable of thermogenesis. In fact, as long as an organism is alive (and actually for a short decay period after its death) and energy is transformed from one form to the other, part of this energy will be dissipated as heat (see Kleiber 1961). Therefore, amphibians and reptiles do generate heat all the time, but as their metabolic rate is normally orders of magnitude lower than a typical endotherm, the amount of heat metabolically derived is, in most instances, negligible for the regulation of their body temperature (Pough 1983). That is, under ordinary conditions, the body temperature of amphibians and reptiles is largely dictated by the availability of thermal niches in the environment, which are explored by behavioral thermoregulation and a few physiological adjustments. However, as far back as 1832, we have been informed by a report from Lamarre-Picquot to the French Academy of Sciences that some python species coil around their eggs while brooding and endogenously elevate their body temperature, possibly for the sake of providing a more propitious environment for the developing embryos. Although the report of Lamarre-Picquot was largely discredited by the French Academy, numerous subsequent studies confirmed and expanded his pioneering observations (see Benedict et al. 1932; Hutchison et al. 1966; Brashears and DeNardo 2013). Since then, the case of brooding pythons incubating their eggs by means of metabolically derived heat (see Figure 6.4) has become "the rule to illustrate the exception" that even organisms belonging to a group readily recognized as ectotherms can, under certain circumstances, exhibit thermogenesis of a magnitude great enough to affect body temperature (Bartholomew 1982). As for the matter considered in this chapter, this adaptation means, once again, that we cannot assume, by default, that the body temperature of reptiles will always be equal to that of the environment.

Although no case of significant thermogenesis has ever being reported for amphibians, there are a few other examples in reptiles. Recently, Casey et al. (2014) showed that leatherback turtles, *Dermochelys coriacea*, can keep their body temperature up to 10°C above water temperature, but this depends greatly on endogenous heat production, which, in turn, requires metabolic rates estimated to be approximately 3 times greater than resting metabolic rates. Tattersall et al. (2004) used infrared thermography to follow heat production in digesting rattlesnakes, *C. durissus*. Like many other snake species, *C. durissus* experience a massive increase in metabolism while digesting their food and, as a thermodynamic side effect, generate enough heat to impact significantly their body temperature (Figure 6.4). Thus, it means that those interested in the effects of temperature on the postprandial metabolic response and energetics of snake digestion cannot assume, as I myself did in the past (Andrade et al. 1997; Toledo et al. 2003; Gavira and Andrade 2013b), that the body temperature of the experimental animals will remain constant and equal to ambient. In fact, as early as 1932, Benedict in his classical book on the physiology of large reptiles mentioned that "as a result of digestion not only is the snake's [python] rectal temperature above the environmental temperature, but likewise its skin temperature" (Benedict 1932).

If brooding pythons and digesting snakes can be regarded as odd examples of transient bouts of thermogenic activity, only relevant for studies focusing on such phenomena, the seasonal thermogenesis by tegu lizards, *S. merianae*, cannot. This lizard has recently been shown to use

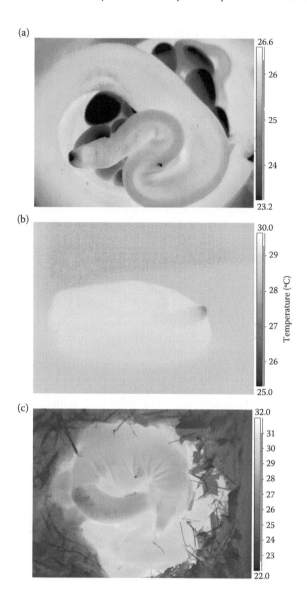

Figure 6.4 Thermogenesis affects the body temperature of some reptile species under different circumstances as revealed by infrared imaging technology (see text for details). (a) Depicts a brooding python, *Python bivittatus*, coiled around her recently laid eggs. (Courtesy of R.S.B. Gavira.) (b) A rattlesnake, *C. durissus*, approximately 24 h after been fed with a rat meal equaling to 20% of its own body mass. (Courtesy of G.J. Tattersall; see also Tattersall et al. 2004.) (c) Depicts an adult tegu lizard, *S. merianae*, whose body temperature is still higher than ambient even after the animal has spent the night retreated in its burrow. (Courtesy of G.J. Tattersall; see Tattersall et al. 2016.)

metabolically derived heat production for a finely tuned body temperature control that enables them to avoid an excessive nightly body temperature drop during the reproductive season (Tattersall et al. 2016). Tegus are heliothermic lizards meaning that they have to dedicate part of their daytime hours to bask in order to elevate body temperature, a thermoregulatory behavior very common for lizards (Huey 1982). Of course, the time an animal has to invest in basking each morning will depend on the initial body temperature in which he engages on this behavior. Thus, regulating nighttime body temperature may represent quite an advantage for quickly reaching the activity body temperature (around 35–37°C for this species; Sanders et al. 2015) during daily early morning basking (Huey and Slatkin 1976). As happens in other ectotherms, tegus are capable of remarkable reductions in heart rate and peripheral vascular resistance shortly after they cease their daily activity and, therefore, "save" some of the heat from 1 day (through the night) to the next (Andrade et al. 2004; Milsom et al. 2012). However, heat dissipation calculations showed that changes in conductance alone would not be enough for keeping the nighttime differential observed between the body temperature of tegus and the temperature of the surroundings (as much as 6°C). Also, tegus were capable of maintaining elevated body temperature even when they were prevented from basking for many days in a row, further demonstrating the endogenous origin of the heat affecting body temperature (see Tattersall et al. 2016; Figure 6.4). Therefore, this case illustrates that an unsuspected medium-bodied lizard (adult body mass typically varying from 2 to 4 kg) is able to endogenously elevate its body temperature well above ambient for an extended period of its life cycle. This response, supposed to be linked to changes in thermoregulatory compromises associated with reproduction, has important evolutionary implications (see discussions in Farmer 2000, 2003; Tattersall et al. 2016). For our immediate interest, the tegus' case reiterates the importance of verifying the body temperature of our experimental subjects during metabolic measurements.

In summary, the similarity between ambient temperature and the body temperature of amphibians and reptiles should not be indiscriminately presumed. Since controlling the body temperature of the animals via the control of ambient temperature may be challenging or even unfeasible under certain circumstances, the one "golden rule" here is to monitor, in as much detail and as accurately as possible, the actual body temperature of the animals while they are being measured. This can be conveniently done with the use temperature dataloggers surgically implanted or ingested (or forcibly ingested) by the experimental animals. Ideally, this should be done simultaneously with the metabolic measurements. If this approach is not possible or recommended under a given protocol, at least the expected body temperature of the animals under that specific experimental setup should be verified. Both the actual measured body temperature and the environmental temperature should be reported in any paper dealing with the metabolic effects of temperature.

Concluding remarks

Temperature is reputed as the single physical factor most influential to any living organism and, as such, temperature effects on an enormous diversity of animal functions have kept generations of biologists busy throughout history. In the last few decades, for the sad reason of human-induced changes in the climate of our planet (IPCC 2014), the study of temperature effects over particular biological systems, and whole ecosystems as well, has gained urgent relevance. In this context, theoretical and empirical efforts have been summoned under a conservationist multidisciplinary approach in an attempt to understand and predict potential problems and outcomes related to animal function under a rapidly changing world (e.g., Navas and Otani 2007; Sinervo et al. 2010; Huey et al. 2012; Seebacher et al. 2014; Bozinovic and Pörtner 2015; Deutsch et al. 2015). While many of these problems and outcomes are yet to be determined, studies focusing on the metabolic correlates of temperature on animals, particularly on ectotherms, can provide an integrative denominator irrefutably relevant for animal life, which is energy flux. In this sense, the central goal of the present chapter is to promote discussion on how our measurements on that front can gain more accuracy and relevance.

The cases discussed in this chapter reiterate the importance of incorporating thermal biology onto metabolic determinations for the particular case of amphibians and reptiles and, more generally, to other ectothermic organisms. The potential caveats resulting from the nonappreciation of this issue were illustrated by the mismatch between the choice of experimental temperatures and the thermal biology in tegu lizards. The relevance of considering other aspects of the thermal environment, besides the averaged mean temperature, was approached by discussing the effects of thermal regime on the digestion of snakes. In this context, there is a growing perception that the potential consequences of temperature variability on the stress and performance of organisms that normally experience fluctuating temperature regimes can only be poorly predicted from the extrapolation of studies carried out under thermal conditions dissociated from the thermal biology of the studied organisms (Ketola et al. 2012; Niehaus et al. 2012; Ketola and Saarinen 2015; Kingsolver et al. 2015; Ma et al. 2015; Vázquez et al. 2015). Finally, by showing that the body temperature of amphibians and reptiles can be considerably different from the ambient, I hope to have fostered a more rigorous control and/ or monitoring of body temperature during the execution of metabolic measurements and the incorporation of such information in the resulting publications.

Acknowledgments

I am in debt to my colleagues and students for sharing my mistakes but, most importantly, for always being open to questioning and helping to find ways to get around them. Our many discussions, often happening in front of an experimental setup in the lab, helped mold the views expressed in this chapter. Rodrigo S. B. Gavira and Ailton Fabrício-Neto kindly permitted the use of published and unpublished data and helped with the figures. Glenn J. Tattersall and Rodrigo S. B. Gavira provided valuable and insightful comments on a previous version of the chapter. During the writing of this chapter, I was supported by the National Council for Scientific and Technological Development (CNPq, grant 302045/2012-0) and by the São Paulo Research Foundation (FAPESP, grant 2013/04190-9).

References

Abe, A.S. 1983. Observations on dormancy in tegu lizard, *Tupinambis teguixin* (Reptilia, Teiidae). *Naturalia* 8:135–139.

Abe, A.S. 1995. Estivation in South American amphibians and reptiles. *Brazilian Journal of Medical and Biological Research* 28:1241–1247.

Agustín, C., Pavão, R., Moreira, C.N., Pinto, A.C.B.C.F., Navas, C.A., and M.T. Rodrigues. 2015. Interaction of morphology, thermal physiology and burrowing performance during the evolution of fossoriality in Gymnophthalmini lizards. *Functional Ecology* 29:515–521.

Andrade, D.V. and A.S. Abe. 1999. Gas exchange and ventilation during dormancy in the tegu lizard, *Tupinambis merianae*. *The Journal of Experimental Biology* 202:3677–3685.

Andrade, D.V., Cruz-Neto, A.P., and A.S. Abe. 1997. Meal size and specific dynamic action in the rattlesnake, *Crotalus durissus* (Serpentes, Viperidae). *Herpetologica* 53:485–493.

Andrade, D.V., Milsom, W.K., Brito, S.P., Toledo, L.F., Wang, T., and A.S. Abe. 2008b. Seasonal changes in daily metabolic patterns of tegu lizards (*Tupinambis merianae*) placed in the cold (17°C) and dark. *Physiological and Biochemical Zoology* 81:165–175.

Andrade, D.V., Sanders, C., Milsom, W.K., and A.S. Abe. 2004. Overwintering in tegu Lizards. In Barnes, B.M. and H.V. Carey (eds), *Life in the Cold: Evolution, Mechanisms, Adaptation, and Application. Twelfth International Hibernation Symposium.* Institute of Artic Biology, University of Alaska: Fairbanks, Alaska, pp. 13–22.

Andrade, D.V., Toledo, L.F., Brito, S.P., Milsom, W.K, and A.S. Abe. 2008a. Effects of season, temperature, and body mass on the standard metabolic rate of tegu lizards (*Tupinambis merianae*). *Physiological and Biochemical Zoology* 81:158–164.

Andrews, R.M. and F.H. Pough. 1985. Metabolism of squamate reptiles: Allometric and ecological relationships. *Physiological Zoology* 58(2):214–231.

Angilletta, M.J. Jr., Niewiarowski, P.H., and C.A. Navas. 2002. The evolution of thermal physiology in ectotherms. *Journal of Thermal Biology* 27:249–268.

Angilletta, M.J. 2009. *Thermal Adaptation: A Theoretical and Empirical Synthesis.* Oxford University Press: New York.

Aubret, F. and R. Shine. 2009. Thermal plasticity in young snakes: How will climate change affect the thermoregulatory tactics of ectotherms? *Journal of Experimental Biology* 213:242–248.

Auffenberg, W. 1981. *The Behavioral Ecology of the Komodo Monitor.* University Presses of Florida: Gainesville, Florida.

Avery, R.A. 1982. Field studies of body temperature and thermoregulation. Temperature, physiology, and the ecology of reptiles. In Gans, C. and F.H. Pough (eds), *Biology of the Reptilia*, Vol. 12. Academic Press: New York, pp. 93–166.

Bakken, G.S. and M.J. Angilletta. 2014. How to avoid errors when quantifying thermal environments. *Functional Ecology* 28:96–107.

Bartelt, P.E. and C.R. Peterson. 2005. Physically modeling operative temperatures and evaporation rates in amphibians. *Journal of Thermal Biology* 30(2):93–102.

Bartholomew, G.A. 1982. Physiological control of body temperature. In Gans, C. and Pough, F.H. (eds), *Biology of the Reptilia*, Vol. 12, Physiology C. Physiological Ecology. Academic Press: London, pp. 167–211.

Basson, C.H. and S. Clusella-Trullas. 2015. The behavior-physiology nexus: Behavioral and physiological compensation are relied on to different extents between seasons. *Physiological and Biochemical Zoology*, 88(4):384–394.

Benedict, F.G. 1932. *The Physiology of Large Reptiles with Special Reference to the Heat Production of Snakes, Tortoises, Lizards, and Alligators.* Carnegie Institute Publications: Washington, DC.

Benedict, F.G., Fox, E.L., and V. Coropatchinsky. 1932. The incubating python: A temperature study. *Proceedings of the National Academy of Sciences of the United States of America* 18(2):209–212.

Bennett, A.F. 1978. Activity metabolism of the lower vertebrates. *Annual Review of Physiology* 40:447–469.

Bennett, A.F. 1980. The metabolic foundations of vertebrate behavior. *BioScience* 30(7):452–456.

Bennett, A.F. 1982. The energetics of reptilian activity. In Gans, C. and F.H. Pough (eds), *Biology of the Reptilia*, Vol. 13. Physiology D. Physiological Ecology. Academic Press: New York, pp. 155–199.

Bennett, A.F. 1994. Exercise performance in reptiles. *Advances in Veterinary Sciences and Comparative Medicine* 38B:113–138.

Bennett, A.F. and P. Licht. 1972. Anaerobic metabolism during activity in lizards. *Journal of Comparative Physiology* 81:277–288.

Bennett, A.F. and P. Licht. 1974. Anaerobic metabolism during activity in amphibians. *Comparative and Biochemistry Physiology* 48A:319–327.

Bennett, A.F. and P.H. Harvey. 1987. Active and resting metabolism in birds: Allometry, phylogeny and ecology. *Journal of Zoology* 213:327–363.

Bennett, A.F. and W.R. Dawson. 1976. Metabolism. In Gans, C. and W.R. Dawson (eds), *Biology of the Reptilia*, Vol. 5. Physiology A. Academic Press: New York, pp. 127–223.

Bostrom, B.L., Jones, T.T., Hastings, M., and D.R. Jones. 2010. Behaviour and physiology: The thermal strategy of leatherback turtles. *PLoS One* 5(11):e13925.

Bozinovic, F. and H.O. Pörtner, 2015. Physiological ecology meets climate change. *Ecology and Evolution* 5(5):1025–1030.

Brashears, J.A. and D.F. DeNardo. 2013. Revisiting python thermogenesis: Brooding Burmese pythons (*Python bivittatus*) cue on body, not clutch, temperature. *Journal of Herpetology* 47(3):440–444.

Brattstrom, B.H. 1963. Preliminary review of the thermal requirements of amphibians. *Ecology* 44:238–255.

Brattstrom, B.H. 1965. Body temperature of reptiles. *The American Midland Naturalist* 73(2):376–422.

Buckley, L.B., Ehrenberger, J.C., and M.J. Angilletta. 2015. Thermoregulatory behavior limits local adaptation of thermal niches and confers sensitivity to climate change. *Functional Ecology* 29:1038–1047.

Caillon, R., Suppo, C., Casas, J., Woods, H.A., and S. Pincebourde. 2014. Warming decreases thermal heterogeneity of leaf surfaces: Implications for behavioural thermoregulation by arthropods. *Functional Ecology* 28:1449–1458.

Caldwell, A.J., While, G.M., Beeon, N.J., and E. Wapsta. 2015. Potential for thermal tolerance to mediate climate change effects on three members of a cool temperate lizard genus, *Niveoscincus*. *Journal of Thermal Biology* 52:14–23.

Carey, C. 2005. How physiological methods and concepts can be useful in conservation biology. *Integrative and Comparative Biology* 45(1):4–11.

Carvalho, J.E., Navas C.A., and I.C. Pereira. 2010. Energy and water in aestivating amphibians. In Navas, C.A. and J.E. Carvalho (eds), *Aestivation: Molecular and Physiological Aspects*. Springer-Verlag: Berlin, pp. 141–169.

Casey, J.P., James, M.C., and A.S. Williard. 2014. Behavioral and metabolic contributions to thermoregulation in freely swimming leatherback turtles at high latitudes. *The Journal of Experimental Biology* 217:2331–2337.

Congdon, J.D., Dnhamn, A.E., and D.W. Tinkle. 1982. Energy budgets and life history of reptiles. In Gans, C. and F.H. Pough (eds), *Biology of the Reptilia, Volume 13. Physiology D. Physiological Ecology*. Academic Press: New York, pp. 233–271.

Dabruzzi, T.F., Sutton, M.A., and W.A. Bennett. 2012. Metabolic thermal sensitivity optimizes sea krait amphibious physiology. *Herpetologica* 68(2):218–225.

Davies, S.J., McGeoch, M.A., and S. Clusella-Trullas. 2015. Plasticity of thermal tolerance and metabolism but not water loss in an invasive reed frog. *Comparative Biochemistry Physiology* 189:11–20.

Deutsch, C., Ferrel, A., Seibel, B., Pörtner, H.O., and R.B. Huey. 2015. Climate change tightens a metabolic constraint on marine habitats. *Science* 348:1132–1135.

di Prampero, P.E. 1985. Metabolic and circulatory limitations to $\dot{V}O_{2max}$ at the whole animal level. *Journal of Experimental Biology* 115:319–331.

Dowd, W.W., King, F.A., and M.W. Denny. 2015. Thermal variation, thermal extremes and the physiological performance of individuals. *Journal of Experimental Biology* 218:1956–1967.

Else, P.L. and A.J. Hulbert. 1981. Comparison of the "mammal machine" and the "reptile machine": Energy production. *American Journal of Physiology* 240:R3–R9.

Else, P.L. and A.J. Hulbert. 1985. An allometric comparison of the mitochondria of mammalian and reptilian tissues: The implications for the evolution of endothermy. *Journal of Comparative Physiology* 156:3–11.

Farmer, C.G. 2000. Parental care: The key to understanding endothermy and other convergent features in birds and mammals. *American Naturalist* 155(3):326–334.

Farmer, C.G. 2003. Reproduction: The adaptive significance of endothermy. *American Naturalist* 162(6):826–840.

Feder M.E. and S.J. Arnold. 1982. Anaerobic metabolism and behavior during predatory encounters between snakes (*Thamnophis elegans*) and salamanders (*Plethodon jordani*). *Oecologia* 53:93–97.

Fraser, S. and G.C. Grigg. 1984. Control of thermal conductance is insignificant to thermoregulation in small reptiles. *Physiological Zoology* 57(4):392–400.

Fray, W., Ackman, R.G., and N. Mrosovsky. 1972. Body temperature of *Dermochelys coriacea*: Warm turtle from cold water. *Science* 177:791–793.

Gates, D.M. 1980. *Biophysical Ecology*. Dover Publications, Inc.: Mineola, New York.

Gatten, R.E., Miller, K., and R.J. Full. 1992. Energetics at rest and during locomotion. In Feder, M.E. and W.W. Burggreen (eds), *Environmental Physiology of the Amphibians*. The University of Chicago Press: Chicago, USA, pp. 314–377.

Gavira, R.S.B. and D.V. Andrade. 2013a. Temperature and thermal regime effects on the specific dynamic action of *Bothrops alternatus* (Serpentes, Viperidae). *Amphibia-Reptilia* 34:483–491.

Gavira, R.S.B. and D.V. Andrade. 2013b. Meal size effects on the postprandial metabolic response of *Bothrops alternatus* (Serpentes: Viperidae). *Zoologia* 30:291–295.

Gillooly, J.F., Brown, J.H., West, G.B., Savage, V.M., and E.L. Charnov. 2001. Effects of size and temperature on metabolic rate. *Science* 293:2248–2251.

Gleeson, T.T. 1991. Patterns of metabolic recovery from exercise in amphibians and reptiles. *Journal of Experimental Biology* 160:187–207.

Gregory, P.T. 1982. Reptilian hibernation. In Gans, C. and F.H. Pough (eds), *Biology of the Reptilia*, Vol. 13, Physiology D. Physiological Ecology. Academic Press: New York, pp. 53–154.

Gunderson, A.R. and M. Leal. 2015. Patterns of thermal constraint on ectotherm activity. *American Naturalist* 185(5): 653–664.

Halsey, L.G., Mattews, P.G.D., Rezende, E.L., Chauvaud, L., and A.A. Robson. 2015. The interactions between temperature and activity levels in driving metabolic rate: Theory, with empirical validation from constrasting ectotherms. *Oecologia* 177:1117–1129.

Hedrick, M.S., Hancock, T.V., and S.S. Hillman. 2015. Metabolism at the max: How vertebrate organisms respond to physical activity. *Comprehensive Physiology* 5:1677–1703.

Hillman, S.S., Withers, P.C., Drewes, R.C., and S.D. Hillyard. 2009. *Ecological and Environmental Physiology of Amphibians*. Oxford University Press: Oxford, UK.

Horne, C.R., Fuller, W.J., Godley, B.J., Rhodes, K.A., Snape, R., Stokes, K.L., and A.C. Broderick. 2014. The effects of thermal variance on the phenotype of marine turtle offspring. *Physiological and Biochemical Zoology* 87(6):796–804.

Huey, R.B. 1982. Temperature, physiology, and the ecology of reptiles. In Gans, C. and F.H. Pough (eds), *Biology of the Reptilia*, Vol. 12. Academic Press: New York, pp. 25–74.

Huey, R.B. and E.R. Pianka. 1977. Seasonal variation in thermoregulatory behavior and body temperature of diurnal Kalahari lizards. *Ecology* 58:1066–1075.

Huey, R.B. and M. Slatkin. 1976. Cost and benefits of lizard thermoregulation. *Quarterly Review of Biology* 51(3):363–384.

Huey, R.B. and R.D. Stevenson. 1979. Integrating thermal physiology and ecology of ectotherms: A discussion of approaches. *American Zoologist* 19:357–366.

Huey, R.B., Kearney, M.R., Krockenberger, A., Holtum, J.A.M., Jess, M., and S.E. Williams. 2012. Predicting organismal vulnerability to climate warming: Roles of behaviour, physiology and adaptation. *Philadelphia Transactions of the Royal Society B* 367:1665–1679.

Hutchison, V.H. and R.K. Dupré. 1992. Thermoregulation. In Feder, M.E. and W.W. Burggreen (eds), *Environmental Physiology of the Amphibians*. The University of Chicago Press: Chicago, IL, pp. 206–249.

Hutchison, V.H., Dowling, H.G., and A. Vinegar. 1966. Thermoregulation in a brooding female Indian python, *Python molurus bivittatus*. *Science* 151:694–696.

IPCC. 2014. In Field, C.B., Barros, V.R., Dokken, D.J., Mach, K.J., Mastrandrea, M.D., Bilir, T.E., Chatterjee, M. et al. (eds), *Climate Change 2014: Impacts, Adaptation, and Vulnerability. Part A: Global and Sectoral Aspects. Contribution of Working Group II to the Fifth Assessment Report of the Intergovernmental Panel on Climate Change*. Cambridge University Press: Cambridge, UK, New York, p. 1132.

Kearn, P., Cramp, R.L., and C.E. Franklin. 2015. Physiological responses of ectotherms to daily temperature variation. *Journal of Experimental Biology* 218:3068–3076.

Kearney, M. and W.P. Porter. 2004. Mapping the fundamental niche: Physiology, climate and the distribution of a nocturnal lizard. *Ecology* 85(11):3119–3131.

Kearney, M., Shine, R., and W.P. Porter. 2009. The potential for behavioral thermoregulation to buffer "cold-blooded" animals against climate warming. *Proceedings of the National Academy of Sciences* 106:3835–3840.

Ketola, T. and K. Saarinen. 2015. Experimental evolution in fluctuating environments: Tolerance measurements at constant temperatures incorrectly predict the ability to tolerate fluctuating temperatures. *Journal of Evolutionary Biology* 28:800–806.

Ketola, T., Kellerman, V., Kristensen, T.N., and V. Loeschcke. 2012. Constant, cycling, hot and cold thermal environments: Strong effects on mean viability but not on genetic estimates. *Journal of Evolutionary Biology* 25:1209–1215.

Kingsolver, J.G., Higgins J.K., and K.E. Augustine. 2015. Fluctuating temperatures and ectotherm growth: Distinguishing non-linear and time-dependent effects. *Journal of Experimental Biology* 218:2218–2225.

Kleiber, K. 1961. *The Fire of Life: An Introduction to Animal Energetics*. Wiley: New York.

Kleiber, M. 1947. Body size and metabolic rate. *Physiological Review* 27:511–541.

Levy, O., Buckley, L.B., Keitt, T.H., Smith, C.D., Boateng, K.O., Kumar, D.S., and M.J. Angilletta. 2015. Resolving the life cycle alters expected impacts of climate change. *Proceedings of the Royal Society B* 282:20150837.

Lighton, J.R.B. 2008. *Measuring Metabolic Rates: A Manual for Scientists*. Oxford University Press: Oxford, UK.

Lorioux, S., DeNardo, D.F., Gorelick, R., and O. Lourdais. 2012. Maternal influences on early development: Preferred temperature prior to oviposition hastens embryogenesis and enhances offspring traits in the Children's python, *Antaresia childreni*. *Journal of Experimental Biology* 215:1346–1353.

Lowell, B.B. and B.M. Spiegelman. 2000. Towards a molecular understanding of adaptive thermogenesis. *Nature* 404:652–660.

Ma, G., Hoffman, A.A., and C.S. Ma. 2015. Daily temperature extremes play an important role in predicting thermal effects. *Journal of Experimental Biology* 218:2289–2296.

Manenti, T., Sørensen, J.G., Moghadam, N.N., and V. Loeschcke. 2014. Predictability rather than amplitude of temperature fluctuations determines stress resistance in a natural population of *Drosophila simulans*. *Journal of Evolutionary Biology* 27:2113–2122.

McNab, B.K. 2002. *The Physiological Ecology of Vertebrates: A View from Energetics.* Comstock/Cornell University Press: Ithaca, New York.

McNab, B.K. and W. Auffenberg. 1976. Temperature regulation of the komodo dragon, *Varanus komodoensis. Comparative and Biochemistry Physiology* 55A: 345–350.

Méndez-Narváez, J., Flechas, S.V., and A. Amézquita. 2015. Foam nests provide context-dependent thermal insulation to embryos of three leptodactylid frogs. *Physiological and Biochemical Zoology* 88(3):246–253.

Milsom, W.K., Sanders, C., Leite, C.A.C., Abe, A.S., Andrade, D.V., and G.J. Tattersall. 2012. Seasonal changes in thermoregulatory strategies of tegu lizards. In Ruf, T., Bieber, C., Arnold, W., and E. Millesi (eds), *Living in a Seasonal World: Thermoregulatory and Metabolic Adaptations.* Springer-Verlag: Berlin, pp. 317–324.

Nagy, K.A. 1983. Ecological energetics. In Huey, R.B., Pianka, E.R., and T.W. Schoener (eds), *Lizard Ecology.* Harvard University Press: Cambridge, pp. 24–54.

Nagy, K.A., Girard, I.A., and T.K. Brown. 1999. Energetics of free-ranging mammals, reptiles, and birds. *Annual Review of Nutrition* 19:247–277.

Navas, C.A. and L. Otani. 2007. Physiology, environmental change, and anuran conservation. *Phyllomedusa* 6(2):83–103.

Newman, A.E., Edmunds, N.B., Ferraro, S., Heffell, Q., Merrit, G.M., Pakkala, J.J., Schilling, C.R., and S. Schorno. 2015. Using ecology to inform physiology studies: Implications of high population density in the laboratory. *American Journal of Regulatory and Integrative Comparative Physiology* 308:R449–R454.

Nichelmann, M. and B. Tzschentke. 1995. Thermoneutrality: Traditions, problems, alternatives. In Nagasaka, T. and A.S. Milton (eds), *Body Temperature and Metabolism.* IPEC: Tokyo, pp. 77–82.

Niehaus, A.C., Angilletta, M.J., Sears, M.W., Franklin, C.E., and R.S. Wilson. 2012. Predicting the physiological performance of ectotherms in fluctuating thermal environments. *Journal of Experimental Biology* 215:694–701.

Paladino, F.V., O'Connor, M.P., and J.R. Spotila. 1990. Metabolism of leatherback turtles, gigantothermy, and thermoregulation of dinosaurs. *Nature* 344: 858–860.

Pinder, A.W., Storey, K.B., and G.R. Ultsch. 1992. Estivation and hibernation. In Feder, M.E. and W.W. Burggreen (eds), *Environmental Physiology of the Amphibians.* The University of Chicago Press: Chicago, IL, pp. 250–274.

Pough, F.H. 1983. Amphibians and reptiles as low-energy systems. In Aspey, W.P. and S.I. Lustick (eds), *Behavioral Energetics: The Cost of Survival in Vertebrates.* Ohio State University Press: Columbus, Ohio, pp. 141–188.

Pough, F.H., Magnusson, W.E., Ryan, M.J., Wells, K.D., and T.L. Taigen. 1992. Behavioral energetics. In Feder, M.E. and W.W. Burggreen (eds), *Environmental Physiology of the Amphibians.* The University of Chicago Press: Chicago, IL, pp. 395–436.

Pough, F.H., Andrews, R.M., Crump, M.L., Savitzky, A.H., Wells, K.D., and M.C. Brandley. 2015. *Herpetology.* 4th Ed. Sinauer: Sunderland, USA.

Rome, L.C., Stevens, E.D., and H.B. John-Alder. 1992. The influence of temperature and thermal acclimation on physiological function. In Feder, M.E. and W.W. Burggreen (eds), *Environmental Physiology of the Amphibians.* The University of Chicago Press: Chicago, IL, pp. 183–205.

Ruben, J.A. 1976. Aerobic and anaerobic metabolism during activity in snakes. *Journal of Comparative Physiology* 109(2):147–157.

Sanders, C.E., Tattersall, G.J., Reichert, M., Andrade, D.V., Abe, A.S., and W.K. Milsom. 2015. Daily and annual cycles in thermoregulatory behavior and cardio-respiratory physiology of black and white tegu lizards. *Journal of Comparative Physiology* 185(8):905–915.

Schmidt-Nielsen, K. 1984. *Scaling: Why Is Animal Size so Important?* Cambridge University Press: Cambridge.

Sears, M.W. and M.J. Angilletta. 2015. Costs and benefits of thermoregulation revisited: Both the heterogeneity and spatial structure of temperature drive energetic costs. *American Naturalist* 185(4):E94–E102.

Seebacher, F. and R. Shine. 2004. Evaluating thermoregulation in reptiles: The fallacy of the inappropriate applied method. *Physiological Biochemical Zoology* 77(4):688–695.

Seebacher, F., Grigg, G.C., and L.A. Beard. 1999. Crocodiles as dinosaurs: Behavioural thermoregulation in very large ectotherms leads to high and stable body temperatures. *Journal of Experimental Biology* 202:77–86.

Seebacher, F., White, C.R., and C.E. Franklin. 2005. Physiological mechanisms of thermoregulation in reptiles: A review. *Journal of Comparative Physiology* 175:533–541.

Seebacher, F., White, C.R., and C.E. Franklin. 2014. Physiological plasticity increases resilience of ectothermic animals to climate change. *Nature Climate Change* 5:61–66.

Shoemaker, V.H., Baker, M.A., and J.P. Loveridge. 1989. Effect of water balance on thermoregulation in waterproof frogs (*Chiromantis* and *Phyllomedusa*). *Physiological Zoology* 62:133–146.

Shoemaker, V.H., McClanahan, L.L., Withers, P.C., Hillman, S.S., and R.C. Drewes. 1987. Thermoregulatory response to heat in the waterproof frogs *Phyllomedusa* and *Chiromantis*. *Physiological Zoology* 60:365–372.

Sinervo, B., Méndez-de-la-Cruz, F., Miles, D.B., Heulin, B., Bastiaans, E., Cruz, M.V., Lara-Resendiz, R. et al. 2010. Erosion of lizard diversity by climate change and altered thermal niches. *Science* 328:1496–1501.

Spotila, J.R., O'Connor, P.O., and G.S. Bakken. 1992. Biophysics of heat and mass transfer. In Feder, M.E. and W.W. Burggreen (eds), *Environmental Physiology of the Amphibians*. The University of Chicago Press: Chicago, IL, pp. 55–80.

Stahlschmidt, Z.R., Jodrey, A.D., and R.L. Luoma. 2015. Consequences of complex environments: Temperature and energy intake interact to influence growth and metabolic rate. *Comparative Biochemistry Physiology* 187:1–7.

Suarez, R.K. 2012. Energy metabolism. *Comprehensive Physiology* 2:2527–2540.

Sunday, J.M., Bates, A.E., Kearney, M.R., Colwell, R.K., Dulvy, N.K., Longino, J.T., and R.B. Huey. 2014. Thermal-safety margins and the necessity of thermoregulatory behavior across latitude and elevation. *Proceedings of the National Academy of Sciences* 111(15):5610–5615.

Tattersall, G.J., Eterovick, P.C., and D.V. Andrade. 2006. Tribute to R.G. Boutilier: Skin colour and body temperature changes in basking *Bokermannohyla alvarengai* (Bokermann, 1956). *Journal of Experimental Biology* 209:1185–1196.

Tattersall, G.J., Leite, C.A.C., Sanders, C., Cadena, V., Andrade, D.V., Abe, A.S., and W.K. Milsom. 2016. Seasonal reproductive endothermy in tegu lizards. *Science Advances* 2(1):e1500951.

Tattersall, G.J., Milsom, W.K., Abe, A.S., Brito, S.P., and D.V. Andrade. 2004. The thermogenesis of digestion in rattlesnakes. *Journal of Experimental Biology* 207:579–585.

Tattersall, G.J., Siclair, B.J., Withers, P.C., Fields, P.A., Seebacher, F., Cooper, C.E., and S.K. Maloney. 2012. Coping with thermal challenges: Physiological adaptations to environmental temperatures. *Comprehensive Physiology* 2:2151–2202.

Toledo, L.F., Abe, A.S., and D.V. Andrade. 2003. Temperature and meal mass effects on the post-prandial metabolism and energetics in a boid snake. *Physiological Zoology* 76:240–246.

Tracy, C.R. 1975. Water and energy relations of amphibians: Insights from mechanistic modelling. In Gates, D.M. and R. Schmerl (eds), *Perspectives in Biophysical Ecology*. Springer-Verlag: Berlin, pp. 325–346.

Tracy, C.R. 1976. A model of the dynamic exchanges of water and energy between a terrestrial amphibian and its environment. *Ecological Monographs* 46(3):293–326.

Tracy, C.R. 1982. Biophysical modeling in reptilian physiology and ecology. In Gans, C. and F.H. Pough (eds), *Biology of the Reptilia*, Vol. 12. Academic Press: New York, pp. 275–321.

Turriago, J.L., Parra, C.A., and M.H. Bernal. 2015. Upper thermal tolerance in anuran embryos and tadpoles at constant and variable peak temperatures. *Canadian Journal of Zoology* 93:267–272.

Underwood, H. 1984. Endogenous rhythms. In Gans, C. and D. Crews (eds), *Biology of the Reptilia*, Vol. 18, Physiology E. Academic Press: New York, pp. 229–297.

Vázquez, D.P., Gianoli, E., Morris, W.F., and F. Bozinovic. 2015. Ecological and evolutionary impacts of changing climatic variability. *Biological Reviews* doi: 10.1111/brv.12216.

White, C.R., Phillips, N.F., and R.S. Seymour. 2006. The scaling and temperature dependence of vertebrate metabolism. *Biological Letters* 2(1):125–127.

Wikelski, M. and S.J. Cooke. 2006. Conservation physiology. *Trends in Ecology and Evolution* 21(1):38–46.

chapter seven

Physiological ecology and conservation of anuran amphibians

**Carlos A. Navas, Fernando R. Gomes,
and Eleonora Aguiar De Domenico**

Contents

Introduction: A general context

That a combination of multiple sources of environmental disruption threatens faunal biodiversity is nowadays considered undisputable. Currently, hundreds of papers document trends across the world with many specific examples, and reviews on the subject have been published sporadically over the past 15 years (Kappelle et al. 1999, de Chazal and Rounsevell 2009, Feehan et al. 2009, Kannan and James 2009, Vittoz et al. 2013). Amphibians not only conform to global tendencies, but they also became the epitomic lineage to illustrate animal extinction due to anthropic actions (Lips et al. 2005, Sodhi et al. 2008, Wake and Vredenburg 2008). Altered climatic/environmental scenarios may act directly and simultaneously at various levels of organization, including individuals, populations, and even communities; and likely affect fauna in complex manners that include bottom-up and top-down interactions. In this chapter, we emphasize bottom-up effects, that is, those perceivable at an individual level, under the principle that anthropogenic change in climate and environment (hereafter termed ACCE) influence the *ecological performance* of individuals and, in turn, affect population dynamics and demography. Specifically, this chapter focuses on how physiological studies can enhance understanding the relationships between putative drivers of environmental change and amphibian declines.

The link between physiological ecology and conservation is rooted in the notion of individual ecological performance, whereas stress and allostatic capacity are the concepts leading the disciplinary convergence between physiological sciences and conservation (Carey 2005, Wikelski and Cooke 2006, Navas and Otani 2007). Let us consider, for example, three possible responses of populations to ACCE (Miles 1994): (i) migration to unaffected areas (as an individual or a population process), (ii) remaining in affected area with increased levels of stress and concomitant reduction in survival and reproductive success, and (iii) remaining as a stable population via physiological or behavioral adjustments. Note that responses (i) and (ii) have mixed bases on population dynamics and physiology whereas (iii) is supported by physiology and autecology, with a recent plea for stronger integration with animal behavior (Frappell 2007, Cooke et al. 2014). Therefore, in the recent past, authors have promoted integration between physiological sciences and biological conservation (Carey et al. 2010), including additional interactions with landscape ecology (Danielsen et al. 2010, Janin et al. 2011), ecological modeling (Sinervo 2010), and animal behavior (Cooke et al. 2014). However, the growing literature on amphibian declines does not accompany the vigorous trend toward higher integration between physiology and conservation that characterizes broader publication trends (Figure 7.1).

In this chapter, we review some basic principles regarding why physiology, as a mechanism-based science, can improve our ability to understand amphibian declines and enhance actions for conservation. Aspects of amphibian conservation physiology in the contexts of chytridiomycosis and habitat fragmentation have been discussed in some detail elsewhere (Navas and Otani 2007). Here, we focus on three topics: physiological correlates of climate change, chronic stress induced by environmental change, and action of water pollutants. Our goal is not to extensively review these topics, but to discuss the potential of physiological ecology to enhance our understanding of such complex problems.

Ecological performance and physiology: Mechanisms of impact

Several authors have acknowledged the value of understanding mechanisms mediating cause-and-effect relationships facing ACCE in the conservation of fauna (as examples see Visser 2008, Sillero and Tarroso 2010, Folguera et al. 2011), and this view necessarily involves physiological

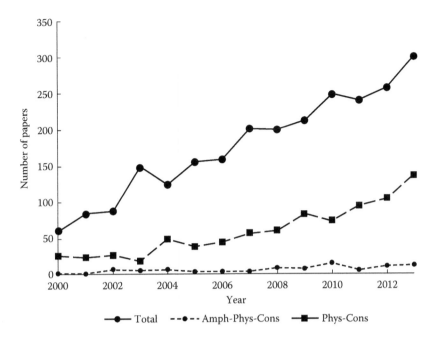

Figure 7.1 Number of papers published in the science citation index (SCI) database covering the topic [amphibian decline] (solid, circles); [physiology and conservation] (traces, square); and [amphibian decline and physiolog*] (short traces, circles).

sciences (see, e.g., Patarnello et al. 2011, Somero 2011, Wharton 2011, Ribeiro et al. 2012). The specific case of amphibian declines requires enhanced hypothetico-deductive research (Li et al. 2013), but inductive research is far more common for conservation concerns that usually start once an ecological pattern has been detected. Thus, a logical step once a pattern has been established is to explore correlational support. This inductive research is valuable and indispensable, yet it is insufficient to produce valid generalizations or to assess the most effective conservation actions for amphibians, even more when the reasons for declines are uncertain. A full conception of problems relating ACCE to amphibian declines requires information on the processes leading to patterns, for example, those mediating shifts in population dynamics (Beebee and Griffiths 2005) or somehow identified as cause-and-effect factors affecting survival or reproductive output (Carey 2005). A possible general path of action from patterns to hypothetico-deductive approaches is illustrated in Figure 7.2, using as a specific example the decline of the natterjack toad analyzed by Beebee (1977). This study was among the first comprehensive attempts to report a pattern of amphibian decline and propose a mechanistic hypothesis linking environmental change to physiological function.

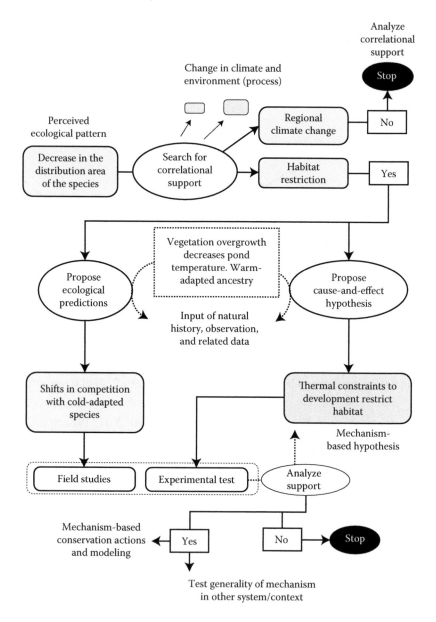

Figure 7.2 Diagram illustrating a possible course of research actions based on the relationship among ecological patterns, processes of environmental degradation, and physiological mechanisms. The proposed course of actions assumes the observation of an ecological pattern as the starting point. The gray boxes exemplify proposed actions with a concrete case based on the work by Beebee on the natterjack toad in Britain. (Adapted from Beebee, T. J. C., 1977. *Biological Conservation* **11**(2): 87–102.)

Physiology, a mechanistic science, has unquestionable value in the design of hypothetico-deductive research of interest in amphibian conservation. The most basic contribution relates to the ultimate goal of comparative physiology, which is understanding how animals work in their environments (Schmidt-Nielsen 1972). Note that this aim is intimately related to the concept of *ecological performance* cited in the above section, which requires understanding which aspects of animal function are more susceptible to environmental change. Ecological performance is essentially an abstraction related to the ability of individuals to survive and reproduce, thus maintaining viable populations in a given environment. In practical research, this abstraction must be substituted by proxies, which often are quantifiable variables based on physiological function and believed to connect survival, growth, behavior, or reproduction to environmental parameters. Accordingly, a fair question is whether our knowledge of amphibian physiology allows proposing the best quantifiable proxies to assess shifts in ecological performance given ACCE and the extent to which selected proxies generalize across lineages and environments. In addition, these proxies are essential for conclusive mechanistic modeling of the impact of ACCE on fauna. For example, the impacts of temperature change on ectothermic vertebrates have produced highly conclusive results on lizards because a well-supported link exists between environmental variables and physiological function (e.g., between environmental temperatures and thermal physiology; Adolph and Porter [1993], Huey et al. [2009], and Sinervo [2010]).

Another contribution of physiology to anuran conservation is based on *reproductive output*. A main link is rooted in the intricate influences of ACCE on the endocrine, immune, and reproductive systems (Cooke et al. 2012), which are consequential at both acute and chronic time frame. Acute effects would be those leading to immediate impaired activity or death, whereas chronic (sublethal) effects would lead to weakened physical condition or decreased reproductive output. These ideas relate to general principles of conservation physiology which have been reviewed elsewhere (Wikelski and Cooke 2006, Cooke et al. 2012, Cooke et al. 2014), including topical papers on amphibians (Navas and Otani 2007, Woods et al. 2010, Blaustein et al. 2012). In addition, physiology contributes to an understanding of why some species are less prone than others to exhibit stress-related reproductive arrest (Wingfield and Sapolsky 2003). Finally, animal reproduction is often intrinsically linked to climate, both directly and through effects on social interactions (Saenz et al. 2006). Thus, direct effects of environment on reproduction may be associated with climate-dependent environmental stimulus, such as photoperiod, rainfall, and temperature. Impacts may have underlying mechanisms at various levels, because stimuli need to be perceived, transduced, and transformed into an organismal response, a sequence that may be disrupted in the context of a stress response. For example, stress can inhibit the hypothalamus–pituitary–gonads axis and affect physiological functions such as vitellogenesis and oocyte growth (Yaron and Siva 2006). Therefore, modified climatic clues can prevent, reduce, or abort reproduction, but can also change the phenology of reproductive events (Todd et al. 2011) or the average fecundity in a population (McCaffery and Maxell 2010).

Amphibian physiological ecology in the context of climate change

The relationship between physiological ecology and climate in anurans has a long-standing history that began with outstanding discussions based on mechanisms (Brattstrom 1968, Beebee 1977). Recent reviews on how climate change affects amphibians have been published elsewhere (Li et al. 2013) and are beyond the scope of this section. Instead, our aim is to highlight aspects of amphibian physiological ecology that are consequential in the context of responses to climate change. This is a main concern for two reasons. One is that amphibians have been understudied from a physiological ecology point of view. A simple search of Web of Science on the topic of physiological ecology illustrates this. A generic search [TOPIC = *Taxon* AND *Factor* AND ecology and physiol*] offers 135 hits when *Taxon* is "lizard" and *Factor* is "temperature," and only 28 if *Taxon* is substituted by "amphibian." One would think that water relationships would have been particularly studied from the standpoint of amphibian physiological ecology, yet this is not the case. The same search using "water" as *Factor* offers 48 hits for "lizard" and 18 for "amphibian." Despite the limitations of such a simple search, clearly we have little information regarding how amphibians work in their environments. However, the Web of Science database shows growing interest in amphibian physiological ecology since 2005, possibly in parallel to a general trend in the field. Under this promising scenario, we highlight some topics relating anuran physiological ecology to climate that remain understudied and that can contribute to amphibian conservation.

Phenology and dormancy: Amphibians usually link reproduction to environmental cues, and therefore may have their phenology impacted by climate change (Beebee 1995, Blaustein et al. 2001, Trumbo et al. 2011). Indeed, a recent review lists phenological shifts among several key nonacute lethal impacts of climate change on amphibians (Li et al. 2013). This situation, when analyzed under the pattern–process–mechanism framed above, calls for immediate attention, for the physiological basis by which those effects take place are largely unknown. Obviously, our predictions regarding the impact of climate change of amphibian phenology would be dramatically different for species directly linking reproduction to rain patterns than to those mainly driven by temperature, and even more distant from those depending on gravitational or geomagnetic cues. Reports exist for all these possibilities. Species linking reproduction to climate may diverge in the dependency of temperature or rain patterns (Saenz et al. 2006), but gravitational or geomagnetic cues are most relevant for some anurans and urodeles and likely enhance changes of synchronized reproduction (Grant et al. 2009). Finally, the timing of phenology can affect the quality of breeding areas (Corn and Muths 2002), thus creating the potential for synergic effects linked to changes in the timing of reproduction.

The physiological processes ending winter dormancy or estivation and the body condition at emergence may be affected by climate, mainly under scenarios of extended dormancy periods (Richter-Boix et al. 2006, Hartel 2008). The body condition of anurans at emergence from dormancy is likely a function of the ability to obtain food and to select proper microhabitats for the inactive season. In addition, the actual dormancy may be more or less challenging according to microclimatic conditions. Variables such as oxygen concentration and duration or magnitude of freezing conditions are essential components of amphibian overwintering performance (Macarthur and Dandy 1982, Tattersall and Ultsch 2008, Swanson and Burdick 2010). Therefore, extended inactivity periods may take a toll on the body condition of both hibernating and estivating amphibians. In this context, where surviving cold conditions may be essential in temperate climates, the maintenance of energy and water balance may be crucial for estivating individuals (Carvalho et al. 2010). Estivation characterizes a significant part of the amphibian fauna in semiarid biomes throughout the world (Carvalho et al. 2010), but the impact of climate change on estivating amphibians remains an elusive topic despite its relevance not only in the context of survival, but also other key aspects of natural history, including growth (Sinsch et al. 2007).

Orientation: In many amphibian species, reproducing individuals must get to reproductive sites and do so by using combined acoustic, celestial, visual, magnetic, or chemical cues that vary in relative importance among lineages (Sinsch 2006). Some anuran species may even discriminate better breeding sites using olfactory cues (Sanuy and Joly 2009). Climate change, then, may affect reproduction not only by altering the climatic cues used to emerge from dormancy, but also those used to find reproductive places. However, predictions regarding the scope and magnitude of impacts will vary according to the dominant aspects of physiology involved in environmental sensing.

Exposure to extreme temperatures during activity: A growing body of literature shows that high critical temperatures are variable among and within lineages and that the risk of overheating varies across environments and lineages (Angilleta 2009). Amphibian larvae may be at various levels of thermal risk according to their physiology, which in turn is influenced by lineage, geographical region, and biome (Duarte et al. 2012).

However, virtually no attention has been given to the impact of near-critical high temperatures in amphibian larvae, even if this type of exposure could elicit stress responses and cause decreased survival or reproductive breakdown (Somero 2011). On the low thermal range, tolerance to freezing certainly determines tolerance to subfreezing winter temperatures prevalent at higher latitudes (Storey and Storey 1996, Costanzo et al. 2008). However, subtropical and high-elevation tropical amphibians may be exposed to frost, even on a daily basis; therefore, climate change has the potential to transform amphibian communities at the threshold for freezing across altitudinal gradients (Carvajalino et al. 2011, Navas et al. 2013). The impact of near-critical temperatures and freezing in tropical and subtropical amphibians remain understudied despite interest from a climate change perspective. Importantly, impacts do not require the actual freezing of whole animals but may involve just a local injury, as it can favor the action of opportunistic microbial pathogens (Nickerson et al. 2011).

Water, temperature, and behavioral performance: In herpetological studies, thermal performance functions have been studied mainly in the context of locomotion capacity, which is a common proxy for fitness. The impact of temperature on locomotion may be consequential in some amphibian lineages, but temperature affects many behavioral and physiological variables. A nonexhaustive list includes retinal perception (Aho et al. 1993), contractile properties of skeletal muscle (Barnes and Ingalls 1991, Girgenrath and Marsh 1997), digestive assimilation efficiency (Benavides et al. 2005), glucose metabolism (Rocha and Branco 1998), performance of the immune system (Raffel et al. 2006), gonad development (Santos et al. 2011), oxygen cost of muscle function (Seebacher et al. 2014), and locomotion (Navas et al. 2008). In addition, the behavioral performance of amphibians is a function of the hydric state of individuals, which is in turn synergic with body temperature (Titon et al. 2010). Thus, simple paths of impacts are likely exceptional, and studies aiming at the simultaneous effects of more than one variable are greatly needed. In anurans, studies combining the impact of water balance and temperature seem indispensable for proper analysis and badly needed for proper mechanistic modeling. In addition, the impact of exposure to unfavorable nonlethal temperatures may be aggravated by decreased trophic resources because the minimum viable time for daily activity would depend on the abundance of trophic resources. Understanding the ecological correlates of metabolic rates in amphibians remains an important topic with new foci emerging in the context of climate change (see Chapter 6).

Physiological adjustment: A main point is that the impact of climate change on amphibians relates to the physiological processes leading to adjustment. Physiological adjustment can be analyzed from many standpoints but is particularly important when asking whether individuals of a given species would be able to adjust to a dimension of ACCE. This matters from many points of view, but we discuss two particularly relevant situations. One is scope for acclimation, which in this case may be considered as the ability of animals to make physiological adjustments so that changed, environment-induced, and negative shifts in physiological function are suppressed or ameliorated (see Chapter 2). Acclimation applies to numerous physiological processes and derived performance functions, but in terms of climate change, the thermal component gains importance. For amphibians, thermal acclimation matters in terms of tolerance to disease (Raffel et al. 2013), adjustment of locomotor performance (Wilson and Franklin 1999), energy flux (Chang and Hou 2005, Rogers et al. 2007), and other highly relevant ecological contexts. Furthermore, tolerance to cold (Zimmerman et al. 2007) and to high temperatures may be rapidly enhanced through acclimatory responses (Brattstrom 1968, Hutchison and Manes 1979). In addition, experimental physiology is necessary to understand the time course and mechanism underlying acclimation. This information is necessary to relate possible physiological rates of adjustment to rates of change in the field, and to establish ecologically relevant thermal limits (Ribeiro et al. 2012). Finally, acclimation matters in several contexts, not only thermal, as many physiological functions are prone to adjustment within given limits. For example, osmotic acclimation exists in anurans (Katz et al. 1984, Katz and Hanke 1993).

To conclude, in the context of anuran conservation, the thermal component of climate has received comparatively more attention than other aspects, yet the influence of sublethal temperatures has been overlooked even if it is intuitively germane to the topic. Physiological ecology has much scope for additional contributions in this context, including entire fields that have received virtually no attention. The physiological ecology of seasonality, for example, is in need of additional studies, particularly in tropical species. The interface between ethology and physiological ecology has enormous potential to generate important contributions in diverse contexts, including orientation and resource searching behaviors.

Physiological ecology of stress in amphibians

Understanding stress responses to anthropogenic change has become an important issue in the convergence between physiology and conservation, since chronic impacts to sublethal conditions are consequential for the stability of animal populations. Changes in ACCE able to produce stress responses in given organisms are known as stressors, compounds that modulate the activity of several physiological control systems. Stressors produce broad organismal impacts through the activation of the sympathetic nervous system and the hypothalamic–pituitary–interrenal axis (HPI axis) (Sapolsky 2002). The activation of the HPI axis culminates in increased secretion of glucocorticoid hormones to the bloodstream, which is related to physiological effects that vary according to the intensity and duration of exposure and to the timing (e.g., bimodal) of response. Generally, short-term activation of the HPI axis temporarily suppresses reproduction and digestion, increases foraging activity and gluconeogenesis, and modulates the immune response. In contrast, long-term activation of the HPI axis typically includes negative functional consequences, for example, chronic suppression of growth, reproduction, and immune function (Sapolsky 2002), all of which are likely important for ecological performance in amphibians. The impact of chronic stressors on physiological function connects directly to population stability and, likely, to bottom-up ecological effects within a community.

The above discussion underlies stress biology and ecoimmunology, two disciplines with main implications and applications to crucial interests in conservation biology. These disciplines have led to promising physiological metrics that can be quantified in individuals and possibly represent valuable proxies of stress conditions in natural populations (Johnstone et al. 2012). Moreover, stress-related physiological variables may anticipate population declines and, consequently, be effective tools in conservation management. For example, this approach has been applied in the context of functional connectivity in amphibian populations exposed to fragmented landscapes (Janin et al. 2012). More generally, global patterns of amphibian decline are due to interacting factors, including anthropogenic stressors such as habitat loss, climate change, invasive species, pollution, and skin pathogens such as the fungus *Batrachochytrium dendrobatidis* (Bd) or iridoviruses from the genus *Ranavirus* (Stuart et al. 2004, Rollins-Smith 2009). Overall, the interaction between stress physiology and ecoimmunology has been relevant for anuran conservation (Carey et al. 1999, Hayes et al. 2010), yet most information on the mechanisms of stress response comes from studies with laboratory mammals, and specific knowledge of amphibian stress biology and immune function is limited.

Reproduction: The impact of chronic stress on amphibian reproduction addresses a most basal aspect of fitness. Both induced stress and exogenous treatments with corticosterone (CORT) can inhibit reproduction, and apparently in various contexts. For example, experimental restraint can increase CORT and decrease androgen plasma levels, this within a few hours from the onset of treatment (Licht et al. 1983, Paolucci et al. 1990, Narayan et al. 2012a,b), and reduce gametogenesis (Pancharatna and Saidapur 1992). Even a common ecological research practice, toe clipping, can decrease plasma levels of testosterone in *Rhinella marina* (Narayan et al. 2011). In addition, treatment with exogenous corticosterone depresses plasma androgen levels (Moore and Zoeller 1985, Paolucci et al. 1990, Narayan et al. 2011), reduces spermatogenesis (Biswas et al. 2000, Tsai et al. 2003), and inhibits reproductive behavior (Moore and Miller 1984, Leary et al. 2006; for a review, see Leary 2009). The inhibition of male courtship behavior by CORT can occur within minutes; therefore, it is likely mediated by nongenomic mechanisms involving signal transduction coupled to membrane glucocorticoid receptors (Orchinik et al. 1991, for a review, see Carr 2011). However, in some species, crowding and captivity can decrease androgen plasma levels without changing CORT (Moore and Zoeller 1985, Houck et al. 1996), similarly to the pattern elicited by close range play-backs of conspecific advertisement calls in treefrogs (Assis et al. 2012, Leary 2014). These findings indicate that the reproductive output of stressed amphibians may be depressed by mechanisms not linked to changes in circulating CORT levels.

An additional complication is that, although many studies have emphasized a negative effect of HPI axis activation on reproduction, natural levels of plasma glucocorticoids may increase during mating in vertebrates (Moore and Jessop 2003, Landys et al. 2006). Then, moderate glucocorticoid plasma levels may facilitate reproduction, possibly by enhancing mobilization of energy substrates (Moore and Jessop 2003, Landys et al. 2006). Such a type of hormonal facilitation of reproduction may characterize anurans as well, mainly because calling is a highly energetic behavior (Wells and Taigen 1989, reviewed in Wells 2001). Anurans display annual and seasonal patterns of circulating CORT (Dupont et al. 1979, Leboulenger et al. 1982, Pancak and Taylor 1983, Narayan et al. 2010), and both CORT and androgen plasma levels can vary even during the reproductive season. This is evident in anurans with prolonged breeding periods, which generally exhibit higher plasma levels of both steroids at the onset of the breeding season (Licht et al. 1983, Mendonca et al. 1985, Assis et al. 2012). Moreover, variation may exist even within males in a chorus, with enhanced levels of androgens and corticosterone in more energetic callers (Orchinik et al. 1988, Emerson and Hess 1996, Harvey et al. 1997, Assis et al. 2012). Understanding how CORT levels vary across natural amphibian life cycles and behavior is crucial to differentiate natural and predictable variation in circulating steroids from variation due to activation of the HPI axis, via responses to acute and chronic stressors.

Stress and immunocompetence: The relationship between CORT and immunocompetence in amphibians has a recent story that already points to a complex picture. Historically, the activation of the HPI axis in vertebrates has been thought to be immunosuppressive (Sapolsky et al. 2000). Pioneering studies emphasized that glucocorticoid administration to amphibians reduces humoral responses and causes lymphopenia, with greater effects on the thymus than in the spleen (Tournefier 1982, Garrido et al. 1987), and inhibits mitogen-stimulated lymphoid proliferation (Highet and Ruben 1987). Consequently, conservation physiology has been influenced by a negative perception of the activation of the HPI axis (Madliger and Love 2014). Anuran responses are complex and generalizations not yet possible. Treatment of frogs (*Rana esculenta*) with pharmacological doses of glucocorticoids inhibits transcription of genes encoding antibacterial skin peptides (Simmaco et al. 1997). However, glucocorticoids may have a role in controlling poststressor responses that are exaggerated and potentially lethal (Sapolsky et al. 2000) and enhance immune responses associated with leukocyte trafficking (Dhabhar and McEwen 1997, 1999). For example, in salamanders (*Cryptobranchus alleganiensis*), restraint time enhances the ability of blood to kill bacteria (bacteria killing ability [BKA]), suggesting that, under some conditions, acute stress improves this immune response (Hopkins and DuRant 2011). Moreover, individuals with patches of discolored skin show higher BKA than individuals with normal integument, suggesting activation of this immune response by a previous bacterial infection (Hopkins and DuRant 2011).

Another relevant issue of amphibian stress biology is that different aspects of the immune response may respond differently to stressors. Cane toads (*R. marina*) decrease BKA in captivity, but individuals maintain typical agglutination of sheep erythrocytes and increase blood phagocytic capacity (Graham et al. 2012). The same authors proposed that the HPI axis possibly inhibits preferentially immune functions that carry risk to individuals, whereas other immune functions may persist or even increase as a stress response. In addition, the duration and intensity of a stressor can modulate immune response in the toad *Rhinella icterica* (Assis et al. 2015). When these toads are kept captive for 24 h, CORT levels triple without apparent changes in indicators of immune response. However, if over the same time a more invasive stress protocol (movement restriction) is applied, toads increase CORT by a ninefold factor, and exhibit altered leukocyte profile and lower plasma BKA. Interestingly, long-term captivity does not mitigate the stress response because captive toads maintained threefold increased CORT even after 3 months of captivity, even if they exhibit a pronounced decline in plasma BKA. Therefore, stress responses and their perceived consequences can be aggravated by time in captivity (Assis et al. 2015). In summary, the overall picture available shows that the effects of stressors and glucocorticoids on the immune response can be dramatically variable within and across species and can be context specific, depending on the intensity and duration of stimuli applied, and that may vary also across the most common variables used as indicators of stress response.

Temperature: Amphibians are ectotherms, and both acute and chronic temperature shifts influence responses to stressors and immunocompetence. Thermal influences seem related to the duration and patterns of thermal variation and to endogenous neuroendocrine rhythms causing season-dependent effects (Zapata et al. 1992). Regarding acute changes, in *R. esculenta*, there is a rise in basal CORT at temperatures above 20°C and, after an adrenocorticotropic hormone (ACTH) injection, CORT levels correlate with temperature between 30°C and 40°C (Jurani et al. 1973). Longer-term thermal shifts have been explored in *R. marina* by acclimating individuals during 14 days to 15°C, 25°C, or 35°C. In such an experiment, both baseline and poststressor urinary CORT are proportional to temperature (Narayan et al. 2012a). In addition, cane toads presented higher corticosterone stress responses when exposed to an acute thermal stressor (Narayan and Hero 2014b) and suppressed physiological endocrine sensitivity to acute stressors when repeatedly exposed to extreme thermal stressors (Narayan and Hero 2014a). Apparently, low temperatures can decrease immunocompetence by reducing the size of lymphoid organs and the synthesis and activity of lymphocytes T and B. These responses occur in parallel to the decreased number of circulating lymphocytes and antibodies and proliferation of lymphocytes T in response to phytohemagglutinin (PHA) and mean serum complement activity (Wright and Cooper 1981, Cooper et al. 1992, Maniero and Carey 1997). After a single injection of bacteriophage F2, cane toads (*R. marina*) exposed during 2–6 weeks at 15°C reduce the appearance of serum antibody in comparison to those maintained at 25°C (Lin and Rowlands 1973). Regarding annual cycles, the toad, *Duttaphrynus melanostictus*, decreases the titer of immunoglobulin IgM and neutrophil counts during hibernation, whereas total white blood cell, eosinophil, and lymphocyte counts remain high (Pratihar and Kundu 2010).

Nutrition: Another important and largely neglected factor is diet quality, which is very likely affected by ACCE via shifts in trophic interactions and consumer-resource dynamics that may influence dietary diversity and prey availability. Tadpoles of *Lithobates sphenocephalus* raised on a low-protein diet reduce PHA and BKA responses in comparison with individuals raised on a high-protein diet (Venesky et al. 2012). Although both temperature and nutritional state can influence amphibian immunocompetence and the stress response, potentially interacting in multiple ways, these subjects remain largely unexplored.

Stress and infection: Exposure to pathogens and parasites elicits stress responses in anurans. For example, *Rana sylvatica* tadpoles exposed to ranaviruses increase baseline CORT and accelerate development, as expected from known influences of this hormone (Warne et al. 2011). Similarly, individuals stimulated to display shorter larval period due to desiccation protocols show comparatively weaker immune responses to PHA challenges and lower leucocyte numbers, a response that becomes stronger with the severity of the protocol (Gervasi and Foufopoulos 2008). Given that chytrid fungus has been strongly associated with global amphibian declines (Fites et al. 2013), the conservation of anurans would be generally promoted by a better understanding of the relationship between disease and immune defenses, and stress response to Bd infection are of particular interest. Free-living males of *Littoria wilcoxii* infected by Bd display comparatively higher baseline urinary CORT (Kindermann et al. 2012). In addition, during a laboratory outbreak of Bd in *Litoria caerulea*, infected and symptomatic specimens increase baseline CORT, decrease plasma potassium and sodium, augment resting metabolic rates, decrease body condition, and exhibit altered white blood cell profiles (Peterson et al. 2013). Interestingly, high dietary protein significantly increased resistance to Bd in tadpoles of *L. sphenocephalus* (Venesky et al. 2012). These results suggest that elevated CORT is associated with relevant infectious diseases and its deleterious effects.

Stress might also interact with macroparasite infection, potentially affecting different aspects of amphibian life history. Toads infected by pulmonary nematodes from the genus *Rhabdias*, for example, show reduced locomotor performance, growth, and survival (Goater and Ward 1992, Goater et al. 1993, Kelehear et al. 2009, Moretti et al. 2014); and males from the treefrog *Hypsiboas prasinus* with high nematode loads are those with lower calling effort (Madelaire et al. 2013). All these phenotypic associations with parasite load might be related to stress responses. Accordingly, individuals of *R. marina* infected by *Rhabdias pseudophaerocephala* show lower CORT responsiveness relative to uninfected toads (Graham et al. 2012). In addition, pesticides may enhance vulnerability to parasites, possibly through suppression of immune responses (see the next section).

Gender and species as additional sources of variation: Besides all sources of variation already cited, males and females may differ in basal CORT levels and may respond differently to stressors. Males of salamanders (*C. alleganiensis*), for example, consistently show higher plasma corticosterone levels than females, a finding congruent with the known territorial activities of males early in the breeding season (Hopkins and DuRant 2011). Similarly, in bullfrogs, the plasma levels of androgens are more sensitive to captivity stress in males than plasma levels of estradiol and testosterone in females (Licht et al. 1983). These variables are linked to expressive interspecific variation, for example, in basal CORT levels, responsiveness to stressors and immunocompetence, a pattern maintained even when phylogenetically close species are compared. Three species of Brazilian toads from the genus *Rhinella*, for example, show interspecific variation in plasma BKA and CORT levels, although they do not differ in the ratio between poststress and baseline levels of CORT. Specifically, mean baseline CORT levels of *Rhinella ornata* are equivalent to those presented by *Rhinella schneideri* after stressful conditions (Gomes et al. 2012). Moreover, BKA and baseline CORT levels are inversely related among species, suggesting that lower innate immunocompetence might be associated with immunosuppressive effects of elevated CORT levels (Gomes et al. 2012). In addition, males from five species of anurans from the Atlantic forest exhibit interspecific variation in BKA during their vocal activity, with mean values varying from 0% to 70% and a pattern of variation that suggests the existence of phylogenetic signal (Assis et al. 2013). Interspecific variation in stress responsiveness and the impact of this variation on homeostasis deserves urgent attention.

Evidences for environmentally induced stress in amphibians: The sources of environmentally induced stress to anurans can be most diverse, including, for example, substrate usage. For example, adult *Bufo bufo* experimentally exposed to substrates that vary according to land use practice avoid ploughed soil and enhance salivary CORT when confined to this soil, in comparison with forest litter or meadows (Janin et al. 2012). Similarly, salamanders (*Ambystoma maculatum*) that migrated over pavement also showed higher plasma CORT than individuals that migrated through undisturbed forest (Homan et al. 2003). Otherwise, salamanders (*C. alleganiensis*) from streams with greater anthropogenic disturbance and from a more forested site show similar baseline and stress-induced CORT and blood BKA (Hopkins and DuRant 2011). Else, habitat availability and fragmentation influence baseline urinary CORT and body condition in toads (*Bufo bufo*) (Janin et al. 2011).

Insights on invasion biology: Another side of the coin is the relationship between stress responses and invasion biology, which matters in conservation, for invasive anurans may be a threat to native fauna. The best example in this context is *R. marina* in Australia. When compared to long-established populations, toads from the invasive front show comparatively lower increase in metabolic rates as a response to inoculated bacterial lipopolysaccharide. This reduced response suggests that selection for increased rates of dispersion might include reduced investment in immune responses (Llewelyn et al. 2010). In addition, relative leg length, a morphological feature associated with invasiveness in *R. marina*, was negatively correlated to the magnitude of CORT response to captivity maintenance, suggesting that the more invasive phenotype is characterized by lower stress response (Graham et al. 2012).

In conclusion, all aspects of stress biology discussed so far suggest a scenario in which diverse indicators of stress seem related to shifts in habitat quality and exposure to disease. In this context, CORT levels can be regarded as a common denominator and its use as an indicator of field stress is certainly relevant for conservation. Given the limited data available, whether or not CORT levels may be a proxy for body condition in anurans remains as an important question. Although much research effort has been applied to investigate amphibian stress response and its physiological consequences, patterns of variation in function of season, general activity, climate, gender, ontogeny, and phylogeny are still poorly covered. This situation hampers scope for generalization and prediction of stress response due to environmental perturbations. The main understudied areas include (1) mechanisms of stress response; (2) dynamics of response to stressors according to intensity and duration; (3) natural patterns of variation of stress response (sexual, ontogenetic, seasonal, populational, and interspecific); and (4) validation of proxies of stress response in the field, and assessment of relevance of populational responses to environmental changes.

Water pollutants and the physiological ecology of larvae

In the last few centuries, human practices have contributed decisively to water quality degradation, resulting in an ubiquitous dispersion of toxicants. A broad variety of man-made chemicals, including fertilizers, pesticides, heavy metals, phenols, hydrocarbons, surfactants, and pharmaceutics, have been added to the environment mainly through runoff from agricultural fields, industrial and domestic sewage, mining drainage, and atmospheric deposition, causing negative impacts on terrestrial and aquatic ecosystems (Forbes and Forbes 1994, Rhind 2009, Newman 2010). This scenario calls for enhanced efforts to understand the fundamental processes underlying the exposition of organisms to environmental toxicants and to convert such studies into better management of chemical discharge (Bueno et al. 2010, Egea-Serrano et al. 2012). This is a key topic in ecotoxicology, a discipline that addresses toxic effects of pollutants on ecosystems, creating tools and theory for assessing the vulnerability of terrestrial and aquatic communities (Van Straalen 2003). Under this framework, the convergence between physiology and conservation biology has become critical to assess ecotoxicological risk, since threats faced by exposed organisms might be better understood by a physiological approach to assess the processes and mechanisms of action for different pollutants (Schulz 2004, Eggen and Suter 2007, Cooke et al. 2013).

At the organismal level, the most important question concerns the patterns of functional impairments and pathologies derived from exposition to pollutants, as well as the underlying mechanisms responsible for such patterns. A diverse array of physiological functions may have potential as a prognostic indicator of altered organismal performance, including those related to the nervous, endocrine, and reproductive system, regulatory enzymes, genetic material, immune response, and stress biology (see previous sections). Such mechanistic approaches apply to amphibians, a taxon for which toxicological studies are still underrepresented, especially with respect to tropical species (Schiesari et al. 2007, Lehman and Williams 2010, Ghose et al. 2014).

Both lethal and sublethal effects of pollutants on amphibians have been reviewed elsewhere (Power et al. 1989, Solomon et al. 2008, Mann et al. 2009, Linder et al. 2010), including efforts in the context of meta-analysis (Kerby et al. 2010, Rohr and Mccoy 2010, Egea-Serrano et al. 2012, Ghose et al. 2014). Important contributions also come from experimental approaches based on microcosm and mesocosm experiments, which have been fundamental to highlighting the interactive effects among pollutants, biotic factors (e.g., food resources, predation, and competition), and physical variables of the environment (Boone et al. 2005, 2007, Davis et al. 2012). Comprehensive reviews on various topics are available and, therefore, our aim in this section is to highlight a group of consequential and observable symptoms caused by exposure to pollutants and to link these symptoms to physiological mechanisms.

Liver injury: In addition to many fundamental functions for anurans, the liver is essential to promote tolerance to cold (Swanson et al. 1996, Edwards et al. 2000), to maintain water balance in face of dehydration (Churchill and Storey 1994), and to provide a glycogen store source for vocal behavior (Wells and Bevier 1997). Contaminants introduced in the aquatic environment may disrupt liver metabolic function in various manners. One known symptom is the decrease of hepatic glycogen content, which has been observed in tadpoles of *Lithobates pipiens* exposed to naphthenic acids (Melvin et al. 2013), but hepatic symptoms may also occur in adult frogs. For example, in *Rana chensinensis*, octylphenol induces acute injury in the liver through upregulation of genes associated with apoptosis, coagulation, innate immunity, and energy metabolism (Li et al. 2014). Similarly, adults of *Xenopus tropicalis* exposed to benzo[*a*]pyrene, a polycyclic aromatic hydrocarbon, exhibit cholesterol accumulation in the liver associated with induction of hepatic apoptosis, necrosis, and autophagy (Regnault et al. 2014). The liver is a central target organ of xenobiotics, thus, studies on the anatomical and physiological state of the amphibian liver may have value for diagnosis, particularly if baseline conditions can be proposed unambiguously. In addition, the links between hepatotoxicity and ecological performance are not clear for many relevant contexts, for example, how this interaction may affect seasonal cycles of activity and influence the energetic sources available for reproduction.

Endocrine disruption: Anuran reproduction depends on a complex chain of endocrine events that influence mating behavior and ultimately fitness (Wilczynski and Lynch 2011). However, hormonal homeostasis during reproduction can be affected by pesticides and by several synthetic substances used as solvents, lubricants, plasticizers, and cosmetics, such as polychlorinated and polybrominated biphenyls, bisphenol-A, phthalates, dioxins, and phytoestrogens, all of which may act as endocrine disruptors leading to reproductive-related symptoms including subfertility (Sifakis and Tsatsakis 2011). This affects both males and females. For example, atrazine, one of the most commonly used pesticides in the world, may cause testicular dysfunctions and hermaphroditism in male amphibians, decreasing androgens, mating behavior, and fertility (Hayes et al. 2002, Tavera-Mendoza et al. 2002, Carr et al. 2003, Langlois et al. 2010, Hayes et al. 2011).

Sexual differentiation is also susceptible to hormonal interference, making aquatic tadpoles especially vulnerable to endocrine disruptors.

However, adult frogs and toads may also absorb substances through the skin and develop gonadal dysfunction even in terrestrial habitats. In several populations of cricket frogs (*Acris crepitans*) studied across field sites in Illinois (USA), for example, a striking relation between the prevalence of intersexuality and the scope of environmental contamination with atrazine, polychlorinated biphenyl, and polychlorinated dibenzofuran has been observed (Reeder et al. 1998). On a related topic, male frogs of *L. pipiens* exposed to a pharmaceutical compound, diethylstilbestrol, synthesize and secrete vitellogenin (Selcer and Verbanic 2014), a major precursor to the egg-yolk proteins of oviparous vertebrates. Moreover, aromatase, an enzyme widely accepted as a key target in endocrine disruption, increases in the brain of both male and female adults of *Xenopus laevis* treated with both water of the Lambro, a highly polluted river in North Italy, and flutamide, a pharmaceutical substance (Massari et al. 2010). These data provide evidence that experimental conditions and realistic scenarios of contamination may lead to pervasive hormonal alterations of typical physiological paths with important impacts on reproduction.

Neurological dysfunction: Behavioral effects of pollutants may influence ecological performance, depressing, accelerating, or somehow impairing normal behavior, which can threaten survival and decrease fitness. Neural functions mediating behavior can be disrupted by organophosphorus pesticides (e.g., malation, chlorpyrifos, and diazinon), an effect likely mediated through binding with cholinesterase (ChE) (Galloway and Handy 2003). Depressed ChE alters amphibian behavior, for example, reducing mobility and producing uncoordinated swimming, two symptoms that presumably enhance vulnerability to predators and mortality and depress growth rates (Bridges 1997, Mariel Aronzon et al. 2014). In contrast, contaminants such as glufosinate ammonium, an herbicide, may increase ChE activity in ways that intensify swimming speed (Peltzer et al. 2013). Compounds such as polychlorinated biphenyls may also reduce tadpole survival through diverse effects on behavior, reducing swimming speed, inducing movement over shorter distances and nonlinear swimming trajectories, or stimulating generalized lethargy in tadpoles (Quimby et al. 2005). Thus, a number of toxic products may affect ecological performance through impacts on behavior which are better documented for larval than for adult phases.

Reduced growth and smaller size at metamorphosis: Many studies have reported impaired growth and delayed metamorphosis in response to toxicant exposure (Diana et al. 2000, Boone and James 2003, Cauble and Wagner 2005, Brodeur et al. 2011). These responses may have partial explanations in hepatotoxicity and endocrine disruption, as previously discussed, but possibly include tandem effects encompassing inhibition of thyroid hormones (Mann et al. 2009), reduced feeding (Bridges 1999), and augmented energetic cost of detoxification (Venturino et al. 2003). Such effects may lead to a reduction in body size at metamorphosis and a consequent decrease in survival rate, particularly during the following terrestrial stage (Altwegg and Reyer 2003). For example, smaller metamorphs have lower locomotion capacity, a trait that may affect toadlet survival (Navas et al. 2007).

Genotoxicity: Genotoxicity relates to the potential effects of toxicants to the integrity of the genetic material, with possible damage to specific genes, that may prevent accurate replication (Yadav et al. 2013). Studies analyzing genotoxicity in amphibians are rare, although the potential of xenobiotics to damage amphibian DNA has been reported by some authors (e.g., Clements et al. 1997, Lajmanovich et al. 2005, Chang et al. 2009, Giri et al. 2012, Singha et al. 2014). This is possibly a very relevant area of research, but generalizations cannot be stated with current information.

Increased susceptibility to disease and parasitism through immunosuppression: Contaminants may reduce immunocompetence in amphibians and facilitate parasite infection (Kiesecker 2002, Christin et al. 2003, Gendron et al. 2003, Forson and Storfer 2006). For example, some studies demonstrate that exposure to pesticides, either alone or mixed with other compounds, accelerates infection and increases nematode and trematode parasite loads (Gendron et al. 1997, Rohr et al. 2008). The mechanisms underlying immunosuppression may include decreased T lymphocyte proliferation and disturbances of the corticosterone balance (Szuroczki and Richardson 2009) (see previous section). These interactions may have relevant consequences for population stability because the simultaneous exposure to water pollutants and diseases illustrates a specific case of competing homeostatic and physiological demands that could affect infection susceptibility and mortality rates.

Biological stress: Several studies point out clearly that contaminants affect stress responses in amphibians. Mudpuppies (*Necturus maculosus*) collected from sites contaminated with pesticides showed reduced CORT response to stressors when compared to individuals collected from reference sites (Gendron et al. 1997). Male toads (*Anaxyrus terrestris*) collected in coal ash-polluted sites also showed higher CORT than males collected in reference sites (Hopkins et al. 1997) and did not respond to an ACTH challenge (Hopkins et al. 1999). Moreover, when individuals of *A. terrestris* from reference sites were transferred to the coal ash-polluted site, CORT increased 10 days posttransference (Hopkins et al. 1997), and CORT increase was not paralleled by corticosteroid-biding globulin (CBG) levels, suggesting that CBG may not provide a protective mechanism during long-term pollution exposure (Ward et al. 2007).

Deformities: Malformations apparently caused by toxicants have been reported mainly over the last decade (Blaustein and Johnson 2003, Linzey et al. 2003, Taylor et al. 2005), also with the potential for synergic interactions. For example, the interaction between pesticides and trematode infection enhances the incidence of deformities in amphibians

(Kiesecker 2002). The increased infestation with trematodes as a response to toxicants reinforces the issues discussed previously, but highlights the diversity of symptoms of parasitic infections, which may include limb malformations and supernumerary limbs (Ankley et al. 2004, Mann et al. 2009). Abnormalities could represent a sublethal response to pesticides and nonagricultural pollutants, such as substances derived from coal combustion, metals, petroleum hydrocarbons, and ammonia (Johnson et al. 2010). However, UV-B radiation and injury from predators are also hypothetical agents that can have additive effects with xenobiotics exposure leading to an increased occurrence of malformations. Thus, more data are necessary to elucidate causative agents in sites where amphibian deformities prevail (Lunde and Johnson 2012).

Concluding, the interface between amphibian physiology and ecotoxicology has contributed to elucidating, among other occurences, the impacts and threats of xenobiotics and to advancing the possible use of amphibian populations for biomonitoring purposes (Carey and Bryant 1995, Zhelev et al. 2013). Within this promising scenario, the main limitations are that most studies are focused on amphibians from temperate regions and, as a consequence, information on tropical species is scarce even for commonly used pesticides (Lehman and Williams 2010, Ghose et al. 2014). In addition, physiological ecology and ecotoxicology can converge on overlooked questions regarding the impact of pollutants on, for example, energetic states, capacity for activity, and reproduction; the implications of changes in the temporal course of metamorphosis; the consequences of pollutant-induced stress; and the impact on performance caused by altered phenotypes.

Concluding remarks

Whereas habitat loss is a major driver of amphibian decline, changes in biotic and abiotic factors can affect ecological performance directly or through influences on how individuals relate to biotic factors. In this context, physiological ecology has enormous potential to grow as a discipline that contributes to amphibian conservation. Several key aspects of anuran physiological ecology have been understudied, and much needed generalizations may be inappropriate given the overrepresentation of temperate species in many crucial contexts. Climate change and exposure to pollutants are two main scenarios for research, where stress physiology emerges as a transversal axis to most types of ACCE.

References

Adolph, S. C. and W. P. Porter, 1993. Temperature, activity, and lizard life histories. *American Naturalist* **142**(2): 273–295.

Aho, A. C., K. Donner, and T. Reuter, 1993. Retinal origins of the temperature effect on absolute visual sensitivity in frogs. *Journal of Physiology* **463**: 501–521.

Altwegg, R. and H. Reyer, 2003. Patterns of natural selection on size at metamorphosis in water frogs. *Evolution* **57**(4): 872–882.

Angilleta, M. J., 2009. *Thermal Adaptation: A Theoretical and Empirical Synthesis.* Oxford, Oxford University Press.

Ankley, G., S. Degitz, S. Diamond, and J. Tietge, 2004. Assessment of environmental stressors potentially responsible for malformations in North American anuran amphibians. *Ecotoxicology and Environmental Safety* **58**(1): 7–16.

Assis, V.R., S.C.M. Titon, A.M.G. Barsotti, B. Titon Jr., F.R. Gomes, 2015. Effects of acute restraint stress, prolonged captivity stress and transdermal corticosterone application on immunocompetence and plasma levels of corticosterone on the cururu toad (*Rhinella icterica*). *PLoS One* **10**(4): e0121005. Doi: 10.1371/journal.pone.0121005.

Assis, V. R., C. A. Navas, M. T. Mendonça, and F. R. Gomes, 2012. Vocal and territorial behavior in the Smith frog (*Hypsiboas faber*): Relationships with plasma levels of corticosterone and testosterone. *Comparative Biochemistry and Physiology* **163A**(3–4): 265–271.

Assis, V. R., S. C. M. Titon, A. M. G. Barsotti, B. Spyra, and F. R. Gomes, 2013. Antimicrobial capacity of plasma from anurans of the Atlantic forest. *South American Journal of Herpetology* **8**(3): 155–160.

Barnes, W. S. and C. P. Ingalls, 1991. Differential effects of temperature on contractile behavior in isolated frog skeletal muscle. *Comparative Biochemistry and Physiology A* **100**: 575–580.

Beebee, T. J. C., 1977. Environmental change as a cause of natterjack toad (*Bufo calamita*) declines in Britain. *Biological Conservation* **11**(2): 87–102.

Beebee, T. J. C., 1995. Amphibian breeding and climate. *Nature* **374**(6519): 219–220.

Beebee, T. J. C. and R. A. Griffiths, 2005. The amphibian decline crisis: A watershed for conservation biology? *Biological Conservation* **125**(3): 271–285.

Benavides, A. G., A. Veloso, P. Jimenez, and M. A. Mendez, 2005. Assimilation efficiency in *Bufo spinulosus* tadpoles (Anura: Bufonidae): Effects of temperature, diet quality and geographic origin. *Revista Chilena De Historia Natural* **78**(2): 295–302.

Biswas, N. M., Chaudhuri, G. R., Sarkar, M., and Sengupta, R. 2000. Influence of adrenal cortex on testicular activity in the toad during the breeding season. *Life Sciences* **66**: 1253–1260.

Blaustein, A. and P. Johnson, 2003. The complexity of deformed amphibians. *Frontiers in Ecology and the Environment* **1**(2): 87–94.

Blaustein, A. R., L. K. Belden, D. H. Olson, D. M. Green, T. L. Root, and J. M. Kiesecker, 2001. Amphibian breeding and climate change. *Conservation Biology* **15**(6): 1804–1809.

Blaustein, A. R., S. S. Gervasi, P. T. J. Johnson, J. T. Hoverman, L. K. Belden, P. W. Bradley, and G. Y. Xie, 2012. Ecophysiology meets conservation: Understanding the role of disease in amphibian population declines. *Philosophical Transactions of the Royal Society B—Biological Sciences* **367**(1596): 1688–1707.

Boone, M., C. Bridges, J. Fairchild, and E. Little, 2005. Multiple sublethal chemicals negatively affect tadpoles of the green frog, *Rana clamitans. Environmental Toxicology and Chemistry* **24**(5): 1267–1272.

Boone, M. and S. James, 2003. Interactions of an insecticide, herbicide, and natural stressors in amphibian community mesocosms. *Ecological Applications* 13(3): 829–841.

Boone, M., R. Semlitsch, E. Little, and M. Doyle, 2007. Multiple stressors in amphibian communities: Effects of chemical contamination, bullfrogs, and fish. *Ecological Applications* 17(1): 291–301.

Brattstrom, B. H., 1968. Thermal acclimation in anuran amphibians as a function of latitude and altitude. *Comparative Biochemistry and Physiology* 24: 93–111.

Bridges, C., 1997. Tadpole swimming performance and activity affected by acute exposure to sublethal levels of carbaryl. *Environmental Toxicology and Chemistry* 16(9): 1935–1939.

Bridges, C., 1999. Effects of a pesticide on tadpole activity and predator avoidance behavior. *Journal of Herpetology* 33(2): 303–306.

Brodeur, J., R. Suarez, G. Natale, A. Ronco, and M. Zaccagnini, 2011. Reduced body condition and enzymatic alterations in frogs inhabiting intensive crop production areas. *Ecotoxicology and Environmental Safety* 74(5): 1370–1380.

Bueno, M., M. Hernando, S. Herrera, M. Gomez, A. Fernandez-Alba, I. Bustamante, and E. Garcia-Calvo, 2010. Pilot survey of chemical contaminants from industrial and human activities in river waters of Spain. *International Journal of Environmental Analytical Chemistry* 90(3–6): 321–343.

Carey, C., 2005. How physiological methods and concepts can be useful in conservation biology. *Integrative and Comparative Biology* 45(1): 4–11.

Carey, C. and C. J. Bryant, 1995. Possible interrelations among environmental toxicants, amphibian development, and decline of amphibian populations. *Environmental Health Perspectives* 103: 13–17.

Carey, C., H. M. Bustamante, and L. J. Livo, 2010. Effects of temperature and hydric environment on survival of the Panamanian golden frog infected with a pathogenic chytrid fungus. *Integrative Zoology* 5(2): 143–153.

Carey, C., N. Cohen, and L. Rollins-Smith, 1999. Amphibian declines: An immunological perspective. *Developmental and Comparative Immunology* 23(6): 459–472.

Carr, J., A. Gentles, E. Smith, W. Goleman, L. Urquidi, K. Thuett, R. Kendall et al., 2003. Response of larval *Xenopus laevis* to atrazine: Assessment of growth, metamorphosis, and gonadal and laryngeal morphology. *Environmental Toxicology and Chemistry* 22(2): 396–405.

Carr, J. A., 2011. Stress and reproduction in amphibians. In Norris D. O. and K. H. Lopez (eds), *Hormones and Reproduction of Vertebrates, Volume 2 Amphibians.* London, Academic Press: 99–116.

Carvajalino, J. M., M. A. Bonilla, and C. A. Navas, 2011. Freezing risk in tropical high-elevation anurans: An assessment based on the Andean frog *Pristimantis nervicus* (Strobomantidae). *South American Journal of Herpetology* 6(2): 73–78.

Carvalho, J. E., C. A. Navas, and I. C. Pereira, 2010. Energy and water in aestivating amphibians. In Navas C. A. and J. E. Carvalho (eds), *Aestivation: Molecular and Physiological Aspects.* Berlin, Springer-Verlag, 49: 141–169.

Cauble, K. and R. Wagner, 2005. Sublethal effects of the herbicide glyphosate on amphibian metamorphosis and development. *Bulletin of Environmental Contamination and Toxicology* 75(3): 429–435.

Chang, J., M. Gu, and K. Kim, 2009. Effect of arsenic on p53 mutation and occurrence of teratogenic salamanders: Their potential as ecological indicators for arsenic contamination. *Chemosphere* **75**(7): 948–954.

Chang, Y. M. and P. C. L. Hou, 2005. Thermal acclimation of metabolic rate may be seasonally dependent in the subtropical anuran Latouche's frog (*Rana latouchii*, Boulenger). *Physiological and Biochemical Zoology* **78**(6): 947–955.

Christin, M., A. Gendron, P. Brousseau, L. Menard, D. Marcogliese, D. Cyr, S. Ruby, and M. Fournier, 2003. Effects of agricultural pesticides on the immune system of *Rana pipiens* and on its resistance to parasitic infection. *Environmental Toxicology and Chemistry* **22**(5): 1127–1133.

Churchill, T. A. and K. B. Storey, 1994. Metabolic responses to dehydration by liver of the wood frog, *Rana sylvatica. Canadian Journal of Zoology—Revue Canadienne De Zoologie* **72**(8): 1420–1425.

Clements, C., S. Ralph, and M. Petras, 1997. Genotoxicity of select herbicides in *Rana catesbeiana* tadpoles using the alkaline single-cell gel DNA electrophoresis (comet) assay. *Environmental and Molecular Mutagenesis* **29**(3): 277–288.

Cooke, S. J., D. T. Blumstein, R. Buchholz, T. Caro, E. Fernández-Juricic, C. E. Franklin, J. Metcalfe et al., 2014. Physiology, behavior, and conservation. *Physiological and Biochemical Zoology* **87**(1): 1–14.

Cooke, S. J., S. G. Hinch, M. R. Donaldson, T. D. Clark, E. J. Eliason, G. T. Crossin, G. D. Raby et al., 2012. Conservation physiology in practice: How physiological knowledge has improved our ability to sustainably manage Pacific salmon during up-river migration. *Philosophical Transactions of the Royal Society B— Biological Sciences* **367**(1596): 1757–1769.

Cooke, S. J., L. Sack, C. E. Franklin, A. P. Farrell, J. Beardall, M. Wikelski, and S. L. Chown, 2013. What is conservation physiology? Perspectives on an increasingly integrated and essential science. *Conservation Physiology* **1**(1): 1–23.

Cooper, E. L., R. K. Wright, A. E. Klempau, and C. T. Smith, 1992. Hibernation alters the frog's immune system. *Cryobiology* **29**(5): 616–631.

Corn, P. S. and E. Muths, 2002. Variable breeding phenology affects the exposure of amphibian embryos to ultraviolet radiation. *Ecology* **83**(11): 2958–2963.

Costanzo, J. P., R. E. Lee, and G. R. Ultsch, 2008. Physiological ecology of overwintering in hatchling turtles. *Journal of Experimental Zoology* **309A**(6): 297–379.

Danielsen, F., N. D. Burgess, P. M. Jensen, and K. Pirhofer-Walzl, 2010. Environmental monitoring: The scale and speed of implementation varies according to the degree of people's involvement. *Journal of Applied Ecology* **47**(6): 1166–1168.

Davis, M., J. Purrenhage, and M. Boone, 2012. Elucidating predator–prey interactions using aquatic microcosms: Complex effects of a crayfish predator, vegetation, and atrazine on tadpole survival and behavior. *Journal of Herpetology* **46**(4): 527–534.

de Chazal, J. and M. D. A. Rounsevell, 2009. Land-use and climate change within assessments of biodiversity change: A review. *Global Environmental Change— Human and Policy Dimensions* **19**(2): 306–315.

Dhabhar, F. S. and B. S. McEwen, 1997. Acute stress enhances while chronic stress suppresses cell-mediated immunity *in vivo*: A potential role for leukocyte trafficking. *Brain Behavior and Immunity* **11**(4): 286–306.

Dhabhar, F. S. and B. S. McEwen, 1999. Enhancing versus suppressive effects of stress hormones on skin immune function. *Proceedings of the National Academy of Sciences of the United States of America* **96**(3): 1059–1064.

Diana, S., W. Resetarits, D. Schaeffer, K. Beckmen, and V. Beasley, 2000. Effects of atrazine on amphibian growth and survival in artificial aquatic communities. *Environmental Toxicology and Chemistry* **19**(12): 2961–2967.

Duarte, H., M. Tejedo, M. Katzenberger, F. Marangoni, D. Baldo, J. F. Beltran, D. A. Marti, A. Richter-Boix, and A. Gonzalez-Voyer, 2012. Can amphibians take the heat? Vulnerability to climate warming in subtropical and temperate larval amphibian communities. *Global Change Biology* **18**(2): 412–421.

Dupont, W., P. Bourgeois, A. Reinberg, and R. Vaillant, 1979. Circannual and circadian rhythms in the concentration of corticosterone in the plasma of the edible frog (*Rana esculenta* L). *Journal of Endocrinology* **80**(1): 117–125.

Edwards, J. R., K. L. Koster, and D. L. Swanson, 2000. Time course for cryoprotectant synthesis in the freeze-tolerant chorus frog, *Pseudacris triseriata*. *Comparative Biochemistry and Physiology Part A—Molecular and Integrative Physiology* **125**(3): 367–375.

Egea-Serrano, A., R. Relyea, M. Tejedo, and M. Torralva, 2012. Understanding of the impact of chemicals on amphibians: A meta-analytic review. *Ecology and Evolution* **2**(7): 1382–1397.

Eggen, R. I. and M. J. Suter, 2007. Analytical chemistry and ecotoxicology—Tasks, needs and trends. *Journal of Toxicology and Environmental Health, Part A* **70**(9): 724–726.

Emerson, S. B. and D. L. Hess, 1996. The role of androgens in opportunistic breeding, tropical frogs. *General and Comparative Endocrinology* **103**(2): 220–230.

Feehan, J., M. Harley, and J. van Minnen, 2009. Climate change in Europe. 1. Impact on terrestrial ecosystems and biodiversity. A review (Reprinted). *Agronomy for Sustainable Development* **29**(3): 409–421.

Fites, J. S., J. P. Ramsey, W. M. Holden, S. P. Collier, D. M. Sutherland, L. K. Reinert, A. S. Gayek et al., 2013. The invasive chytrid fungus of amphibians paralyzes lymphocyte responses. *Science* **342**: 366–369.

Folguera, G., D. A. Bastias, J. Caers, J. M. Rojas, M. D. Piulachs, X. Belles, and F. Bozinovic, 2011. An experimental test of the role of environmental temperature variability on ectotherm molecular, physiological and life-history traits: Implications for global warming. *Comparative Biochemistry and Physiology* **159A**(3): 242–246.

Forbes, V. E. and T. L. Forbes, 1994. *Ecotoxicology in Theory and Practice*. London, Chapman & Hall.

Forson, D. and A. Storfer, 2006. Effects of atrazine and iridovirus infection on survival and life-history traits of the long-toed salamander (*Ambystoma macrodactylum*). *Environmental Toxicology and Chemistry* **25**(1): 168–173.

Frappell, P., 2007. Integrating physiology, behaviour and ecology: The diverse approach to a coherent blueprint for conservation. *Comparative Biochemistry and Physiology Part A—Molecular & Integrative Physiology* **146**(4): S81-S82.

Galloway, T. and R. Handy, 2003. Immunotoxicity of organophosphorous pesticides. *Ecotoxicology* **12**(1–4): 345–363.

Garrido, E., R. P. Gomariz, J. Leceta, and A. Zapata, 1987. Effects of dexamethasone on the lymphoid organs of *Rana perezi*. *Developmental & Comparative Immunology* **11**(2): 375–384.

Gendron, A. D., C. A. Bishop, R. Fortin, and A. Hontela, 1997. *In vivo* testing of the functional integrity of the corticosterone-producing axis in mudpuppy (amphibia) exposed to chlorinated hydrocarbons in the wild. *Environmental Toxicology and Chemistry* **16**(8): 1694–1706.

Gendron, A. D., D. J. Marcogliese, S. Barbeau, M. S. Christin, P. Brousseau, S. Ruby, D. Cyr, and M. Fournier, 2003. Exposure of leopard frogs to a pesticide mixture affects life history characteristics of the lungworm *Rhabdias ranae*. *Oecologia* **135**(3): 469–476.

Gervasi, S. S. and J. Foufopoulos, 2008. Costs of plasticity: Responses to desiccation decrease post-metamorphic immune function in a pond-breeding amphibian. *Functional Ecology* **22**(1): 100–108.

Ghose, S. L., M. A. Donnelly, J. Kerby, and S. M. Whitfield, 2014. Acute toxicity tests and meta-analysis identify gaps in tropical ecotoxicology for amphibians. *Environmental Toxicology and Chemistry* **33**(9): 2114–2119.

Girgenrath, M. and R. L. Marsh, 1997. *In vivo* performance of trunk muscles in tree frogs during calling. *Journal of Experimental Biology* **200**: 3101–3108.

Giri, A., S. Yadav, S. Giri, and G. Sharma, 2012. Effect of predator stress and malathion on tadpoles of Indian skittering frog. *Aquatic Toxicology* **106**: 157–163.

Goater, C. P., R. D. Semlitsch, and M. V. Bernasconi, 1993. Effects of body size and parasite infection on the locomotory performance of juvenile toads, *Bufo bufo*. *Oikos* **66**: 129–136.

Goater, C. P. and P. I. Ward, 1992. Negative effects of *Rhabdias bufonis* (Nematoda) on the growth and survival of toads (*Bufo bufo*). *Oecologia* **89**(2): 161–165.

Gomes, F. R., R. V. Oliveira, B. Titon, R. Moretti, and B. Mendonca, 2012. Interspecific variation in innate immune defenses and stress response of toads from Botucatu (São Paulo, Brazil). *South American Journal of Herpetology* **7**: 1–8.

Graham, S. P., C. Kelehear, G. P. Brown, and R. Shine, 2012. Corticosterone-immune interactions during captive stress in invading Australian cane toads (*Rhinella marina*). *Hormones and Behavior* **62**(2): 146–153.

Grant, R. A., E. A. Chadwick, and T. Halliday, 2009. The lunar cycle: A cue for amphibian reproductive phenology? *Animal Behaviour* **78**(2): 349–357.

Hartel, T., 2008. Weather conditions, breeding date and population fluctuation in *Rana dalmatina* from central Romania. *Herpetological Journal* **18**(1): 40–44.

Harvey, L. A., C. R. Propper, S. K. Woodley, and M. C. Moore, 1997. Reproductive endocrinology of the explosively breeding desert spadefoot toad, *Scaphiopus couchii*. *General and Comparative Endocrinology* **105**(1): 102–113.

Hayes, T., K. Haston, M. Tsui, A. Hoang, C. Haeffele, and A. Vonk, 2002. Herbicides: Feminization of male frogs in the wild. *Nature* **419**(6910): 895–896.

Hayes, T. B., L. L. Anderson, V. R. Beasley, S. R. de Solla, T. Iguchi, H. Ingraham, P. Kestemont et al., 2011. Demasculinization and feminization of male gonads by atrazine: Consistent effects across vertebrate classes. *Journal of Steroid Biochemistry and Molecular Biology* **127**(1–2): 64–73.

Hayes, T. B., P. Falso, S. Gallipeau, and M. Stice, 2010. The cause of global amphibian declines: A developmental endocrinologist's perspective. *Journal of Experimental Biology* **213**(6): 921–933.

Highet, A. B. and L. N. Ruben, 1987. Corticosteroid regulation of IL-1 production may be responsible for deficient immune suppressor function during the metamorphosis of *Xenopus laevis*, the South African clawed toad. *Immunopharmacology* **13**(2): 149–155.

Homan, R. N., J. V. Regosin, D. M. Rodrigues, J. M. Reed, B. S. Windmiller, and L. M. Romero, 2003. Impacts of varying habitat quality on the physiological stress of spotted salamanders (*Ambystoma maculatum*). *Animal Conservation* **6**: 11–18.

Hopkins, W. A. and S. E. DuRant, 2011. Innate immunity and stress physiology of eastern hellbenders (*Cryptobranchus alleganiensis*) from two stream reaches with differing habitat quality. *General and Comparative Endocrinology* **174**(2): 107–115.

Hopkins, W. A., M. T. Mendonca, and J. D. Congdon, 1997. Increased circulating levels of testosterone and corticosterone in southern toads, *Bufo terrestris*, exposed to coal combustion waste. *General and Comparative Endocrinology* **108**(2): 237–246.

Hopkins, W. A., M. T. Mendonca, and J. D. Congdon, 1999. Responsiveness of the hypothalamo–pituitary–interrenal axis in an amphibian (*Bufo terrestris*) exposed to coal combustion wastes. *Comparative Biochemistry and Physiology, Part C—Pharmacology Toxicology & Endocrinology* **122**(2): 191–196.

Houck, L. D., M. T. Mendonca, T. K. Lynch, and D. E. Scott, 1996. Courtship behavior and plasma levels of androgens and corticosterone in male marbled salamanders, *Ambystoma opacum* (ambystomatidae). *General and Comparative Endocrinology* **104**(2): 243–252.

Huey, R. B., C. A. Deutsch, J. J. Tewksbury, L. J. Vitt, P. E. Hertz, H. J. A. Perez, and T. Garland, 2009. Why tropical forest lizards are vulnerable to climate warming. *Proceedings of the Royal Society B—Biological Sciences* **276**(1664): 1939–1948.

Hutchison, V. H. and J. D. Manes, 1979. The role of behavior in temperature acclimation and tolerance in ectotherms. *American Zoologist* **19**: 367–384.

Janin, A., J. P. Lena, S. Deblois, and P. Joly, 2012. Use of stress hormone levels and habitat selection to assess functional connectivity of a landscape for an amphibian. *Conservation Biology* **26**(5): 923–931.

Janin, A., J. P. Lena, and P. Joly, 2011. Beyond occurrence: Body condition and stress hormone as integrative indicators of habitat availability and fragmentation in the common toad. *Biological Conservation* **144**(3): 1008–1016.

Johnson, P. T. J., M. K. Reeves, S. K. Krest, and A. E. Pinkney, 2010. A decade of deformities: Advances in our understanding of amphibian malformations and their implications. In Sparling, D. W., G. Linder, C. A. Bishop, and S. K. Krest (eds), *Ecotoxicology of Amphibians and Reptiles*. Pensacola, FL, SETAC Press: 511–536.

Johnstone, C. P., R. D. Reina, and A. Lill, 2012. Interpreting indices of physiological stress in free-living vertebrates. *Journal of Comparative Physiology B—Biochemical Systemic and Environmental Physiology* **182**(7): 861–879.

Jurani, M., K. Murgas, L. Mikulaj, and F. Babusiko, 1973. Effect of stress and environmental temperature on adrenal function in *Rana esculenta*. *Journal of Endocrinology* **57**(3): 385–391.

Kannan, R. and D. A. James, 2009. Effects of climate change on global biodiversity: A review of key literature. *Tropical Ecology* **50**(1): 31–39.

Kappelle, M., M. M. I. Van Vuuren, and P. Baas, 1999. Effects of climate change on biodiversity: A review and identification of key research issues. *Biodiversity and Conservation* **8**(10): 1383–1397.

Katz, U., G. Degani, and S. Gabbay, 1984. Acclimation of the euryhaline toad *Bufo viridis* to hyperosmotic solution (NaCl, urea and mannitol). *Journal of Experimental Biology* **108**(January): 403–409.

Katz, U. and W. Hanke, 1993. Mechanisms of hyperosmotic acclimation in *Xenopus laevis* (salt, urea or mannitol). *Journal of Comparative Physiology* **163B**(3): 189–195.

Kelehear, C., J. K. Webb, and R. Shine, 2009. *Rhabdias pseudosphaerocephala* infection in *Bufo marinus*: Lung nematodes reduce viability of metamorph cane toads. *Parasitology* **136**(8): 919–927.

Kerby, J., K. Richards-Hrdlicka, A. Storfer, and D. Skelly, 2010. An examination of amphibian sensitivity to environmental contaminants: Are amphibians poor canaries? *Ecology Letters* **13**(1): 60–67.

Kiesecker, J., 2002. Synergism between trematode infection and pesticide exposure: A link to amphibian limb deformities in nature? *Proceedings of the National Academy of Sciences of the United States of America* **99**(15): 9900–9904.

Kindermann, C., E. J. Narayan, and J. M. Hero, 2012. Urinary corticosterone metabolites and chytridiomycosis disease prevalence in a free-living population of male Stony Creek frogs (*Litoria wilcoxii*). *Comparative Biochemistry and Physiology Part A—Molecular & Integrative Physiology* **162**(3): 171–176.

Lajmanovich, R., M. Cabagna, P. Peltzer, G. Stringhini, and A. Attademo, 2005. Micronucleus induction in erythrocytes of the *Hyla pulchella* tadpoles (Amphibia: Hylidae) exposed to insecticide endosulfan. *Mutation Research— Genetic Toxicology and Environmental Mutagenesis* **587**(1–2): 67–72.

Landys, M. M., M. Ramenofsky, and J. C. Wingfield, 2006. Actions of glucocorticoids at a seasonal baseline as compared to stress-related levels in the regulation of periodic life processes. *General and Comparative Endocrinology* **148**(2): 132–149.

Langlois, V. S., A. C. Carew, B. D. Pauli, M. G. Wade, G. M. Cooke, and V. L. Trudeau, 2010. Low levels of the herbicide Atrazine alter sex ratios and reduce metamorphic success in *Rana pipiens* tadpoles raised in outdoor mesocosms. *Environmental Health Perspectives* **118**(4): 552–557.

Leary, C. J., 2009. Hormones and acoustic communication in anuran amphibians. *Integrative and Comparative Biology* **49**(4): 452–470.

Leary, C. J., 2014. Male green treefrogs use acoustic signals to manipulate the stress physiology of receivers. *Integrative and Comparative Biology* **54**: E304–E304.

Leary, C. J., A. M. Garcia, and R. Knapp, 2006. Stress hormone is implicated in satellite-caller associations and sexual selection in the Great Plains toad. *American Naturalist* **168**(4): 431–440.

Leboulenger, F., C. Delarue, A. Belanger, I. Perroteau, P. Netchitailo, P. Leroux, S. Jegou, M. C. Tonon, and H. Vaudry, 1982. Direct radioimmunoassays for plasma corticosterone and aldosterone in frog: I. Validation of the methods and evidence for daily rhythms in a natural environment. *General and Comparative Endocrinology* **46**(4): 521–532.

Lehman, C. M. and B. K. Williams, 2010. Effects of current-use pesticides on amphibians. In Sparling, D. W., G. Linder, C. A. Bishop, and S. K. Krest (eds), *Ecotoxicology of Amphibians and Reptiles*. Pensacola, FL, SETAC Press: 551.

Li, X.-Y., N. Xiao, and Y.-H. Zhang, 2014. Toxic effects of octylphenol on the expression of genes in liver identified by suppression subtractive hybridization of *Rana chensinensis. Ecotoxicology* **23**(1): 1–10.

Li, Y. M., J. M. Cohen, and J. R. Rohr, 2013. Review and synthesis of the effects of climate change on amphibians. *Integrative Zoology* **8**(2): 145–161.

Licht, P., B. R. McCreery, R. Barnes, and R. Pang, 1983. Seasonal and stress related changes in plasma gonadotropins, sex steroids, and corticosterone in the bullfrog, *Rana catesbeiana. General and Comparative Endocrinology* **50**(1): 124–145.

Lin, H. H. and D. T. Rowlands, 1973. Thermal regulation of the immune response in South American toads (*Bufo marinus*). *Immunology* 24: 129–133.

Linder, G., C. M. Lehman, and J. R. Bidwell, 2010. *Ecotoxicology of Amphibians and Reptiles in a Nutshell*. New York, Society of Environmental Toxicology and Chemistry (SETAC).

Linzey, D., J. Burroughs, L. Hudson, M. Marini, J. Robertson, J. Bacon, M. Nagarkatti, and P. Nagarkatti, 2003. Role of environmental pollutants on immune functions, parasitic infections and limb malformations in marine toads and whistling frogs from Bermuda. *International Journal of Environmental Health Research* 13(2): 125–148.

Lips, K. R., P. A. Burrowes, J. R. Mendelson, and G. Parra-Olea, 2005. Amphibian declines in Latin America: Widespread population declines, extinctions, and impacts. *Biotropica* 37(2): 163–165.

Llewelyn, J., B. L. Phillips, R. A. Alford, L. Schwarzkopf, and R. Shine, 2010. Locomotor performance in an invasive species: Cane toads from the invasion front have greater endurance, but not speed, compared to conspecifics from a long-colonised area. *Oecologia* 162(2): 343–348.

Lunde, K. B. and P. T. J. Johnson, 2012. A practical guide for the study of malformed amphibians and their causes. *Journal of Herpetology* 46(4): 429–441.

Macarthur, D. L. and J. W. T. Dandy, 1982. Physiological aspects of overwintering in the boreal chorus frog (*Pseudacris triseriata maculata*). *Comparative Biochemistry and Physiology* 72A(1): 137–141.

Madelaire, C. B., R. J. da Silva, and F. R. Gomes, 2013. Calling behavior and parasite intensity in treefrogs, *Hypsiboas prasinus*. *Journal of Herpetology* 47(3): 450–455.

Madliger, C. L. and O. P. Love, 2014. The need for a predictive, context-dependent approach to the application of stress hormones in conservation. *Conservation Biology* 28(1): 283–287.

Maniero, G. D. and C. Carey, 1997. Changes in selected aspects of immune function in the leopard frog, *Rana pipiens*, associated with exposure to cold. *Journal of Comparative Physiology* 167B(4): 256–263.

Mann, R., R. Hyne, C. Choung, and S. Wilson, 2009. Amphibians and agricultural chemicals: Review of the risks in a complex environment. *Environmental Pollution* 157(11): 2903–2927.

Mariel Aronzon, C., D. J. G. Marino, A. E. Ronco, and C. S. Perez Coll, 2014. Differential toxicity and uptake of Diazinon on embryo-larval development of *Rhinella arenarum*. *Chemosphere* 100: 50–56.

Massari, A., R. Urbatzka, A. Cevasco, L. Canesi, C. Lanza, L. Scarabelli, W. Kloas, and A. Mandich, 2010. Aromatase mRNA expression in the brain of adult *Xenopus laevis* exposed to Lambro river water and endocrine disrupting compounds. *General and Comparative Endocrinology* 168(2): 262–268.

McCaffery, R. M. and B. A. Maxell, 2010. Decreased winter severity increases viability of a montane frog population. *Proceedings of the National Academy of Sciences of the United States of America* 107(19): 8644–8649.

Melvin, S., C. Lanctot, P. Craig, T. Moon, K. Peru, J. Headley, and V. Trudeau, 2013. Effects of naphthenic acid exposure on development and liver metabolic processes in anuran tadpoles. *Environmental Pollution* 177: 22–27.

Mendonca, M. T., P. Licht, M. J. Ryan, and R. Barnes, 1985. Changes in hormone levels in relation to breeding behavior in male bullfrogs (*Rana catesbeiana*) at the individual and population levels. *General and Comparative Endocrinology* 58(2): 270–279.

Miles, D. B., 1994. Population differentiation in locomotor performance and the potential response of a terrestrial organism to global environmental change. *American Zoologist* **34**(3): 422–436.

Moore, F. L. and L. J. Miller, 1984. Stress-induced inhibition of sexual behavior: Corticosterone inhibits courtship behaviors of a male amphibian (*Taricha granulosa*). *Hormones and Behavior* **18**(4): 400–410.

Moore, F. L. and R. T. Zoeller, 1985. Stress-induced inhibition of reproduction: Evidence of suppressed secretion of Lh-Rh in an amphibian. *General and Comparative Endocrinology* **60**(2): 252–258.

Moore, I. T. and T. S. Jessop, 2003. Stress, reproduction, and adrenocortical modulation in amphibians and reptiles. *Hormones and Behavior* **43**(1): 39–47.

Moretti, R., C. B. Madelaire, R. J. Silva, M. T. Mendonca, and F. R. Gomes, 2014. The relationships between parasite intensity, locomotor performance, and body condition in adult toads (*Rhinella icterica*) from the wild. *Journal of Herpetology* **48**: 277–283.

Narayan, E. and J. Hero, 2014a. Acute thermal stressor increases glucocorticoid response but minimizes testosterone and locomotor performance in the cane toad (*Rhinella marina*). *PLoS One* **9**(3): e92090. Doi: 10.1371/journal. pone.0092090.

Narayan, E. and J. Hero, 2014b. Repeated thermal stressor causes chronic elevation of baseline corticosterone and suppresses the physiological endocrine sensitivity to acute stressor in the cane toad (*Rhinella marina*). *Journal of Thermal Biology* **41**: 72–76.

Narayan, E., F. Molinia, K. Christi, C. Morley, and J. Cockrem, 2010. Urinary corticosterone metabolite responses to capture, and annual patterns of urinary corticosterone in wild and captive endangered Fijian ground frogs (*Platymantis vitiana*). *Australian Journal of Zoology* **58**(3): 189–197.

Narayan, E. J., J. F. Cockrem, and J. M. Hero, 2011a. Urinary corticosterone metabolite responses to capture and captivity in the cane toad (*Rhinella marina*). *General and Comparative Endocrinology* **173**(2): 371–377.

Narayan, E. J., J. F. Cockrem, and J. M. Hero, 2012a. Effects of temperature on urinary corticosterone metabolite responses to short-term capture and handling stress in the cane toad (*Rhinella marina*). *General and Comparative Endocrinology* **178**: 301–305.

Narayan, E. J., J. F. Cockrem, and J. M. Hero, 2012b. Urinary corticosterone metabolite responses to capture and handling in two closely related species of free-living Fijian frogs. *General and Comparative Endocrinology* **177**(1): 55–61.

Narayan, E. J., F. C. Molinia, C. Kindermann, J. F. Cockrem, and J. M. Hero, 2011b. Urinary corticosterone responses to capture and toe-clipping in the cane toad (*Rhinella marina*) indicate that toe-clipping is a stressor for amphibians. *General and Comparative Endocrinology* **174**(2): 238–245.

Navas, C. A., M. M. Antoniazzi, J. E. Carvalho, H. Suzuki, and C. Jared, 2007. Physiological basis for diurnal activity in dispersing juvenile *Bufo granulosus* in the Caatinga, a Brazilian semi-arid environment. *Comparative Biochemistry and Physiology Part A—Molecular & Integrative Physiology* **147**(3): 647–657.

Navas, C. A., J. M. Carvajalino-Fernández, L. P. Saboyá-Acosta, L. A. Rueda-Solano, and M. A. Carvajalino-Fernández, 2013. The body temperature of active amphibians along a tropical elevation gradient: Patterns of mean and variance and inference from environmental data. *Functional Ecology* **27**: 1145–1154.

Navas, C. A., F. R. Gomes, and J. E. Carvalho, 2008. Thermal relationships and exercise physiology in anuran amphibians: Integration and evolutionary implications. *Comparative Biochemistry and Physiology Part A—Molecular & Integrative Physiology* **151**(3): 344–362.

Navas, C. A. and L. Otani, 2007. Physiology, environmental change, and anuran conservation. *Phyllomedusa* **6**(2): 83–103.

Newman, M. C., 2010. *Fundamentals of Ecotoxicology*. Boca Raton, FL, CRC.

Nickerson, C. A., C. M. Ott, S. L. Castro, V. M. Garcia, T. C. Molina, J. T. Briggler, A. L. Pitt et al., 2011. Evaluation of microorganisms cultured from injured and repressed tissue regeneration sites in endangered giant aquatic Ozark hellbender salamanders. *PLoS One* **6**(12): e28906. Doi: 10.1371/journal.pone.0028906.

Orchinik, M., P. Licht, and D. Crews, 1988. Plasma steroid concentrations change in response to sexual behavior in *Bufo marinus*. *Hormones and Behavior* **22**(3): 338–350.

Orchinik, M., T. F. Murray, and F. L. Moore, 1991. A corticosteroid receptor in neuronal membranes. *Science* **252**(5014): 1848–1851.

Pancak, M. K. and D. H. Taylor, 1983. Seasonal and daily plasma corticosterone rhythms in American toads, *Bufo americanus*. *General and Comparative Endocrinology* **50**(3): 490–497.

Pancharatna, K. and S. K. Saidapur, 1992. A study of ovarian follicular kinetics, oviduct, fat-body, and liver mass cycles in laboratory-maintained *Rana cyanophlyctis* in comparison with wild-caught frogs. *Journal of Morphology* **214**(2): 123–129.

Paolucci, M., V. Esposito, M. M. Difiore, and V. Botte, 1990. Effects of short postcapture confinement on plasma reproductive hormone and corticosterone profiles in *Rana esculenta* during the sexual cycle. *Bollettino di Zoologia* **57**(3): 253–259.

Patarnello, T., C. Verde, G. Di Prisco, L. Bargelloni, and L. Zane, 2011. How will fish that evolved at constant sub-zero temperatures cope with global warming? Notothenioids as a case study. *Bioessays* **33**(4): 260–268.

Peltzer, P. M., C. M. Junges, A. M. Attademo, A. Basso, P. Grenon, and R. C. Lajmanovich, 2013. Cholinesterase activities and behavioral changes in *Hypsiboas pulchellus* (Anura: Hylidae) tadpoles exposed to glufosinate ammonium herbicide. *Ecotoxicology* **22**(7): 1165–1173.

Peterson, J. D., J. E. Steffen, L. K. Reinert, P. A. Cobine, A. Appel, L. Rollins-Smith, and M. T. Mendonca, 2013. Host stress response is important for the pathogenesis of the deadly amphibian disease, chytridiomycosis, in *Litoria caerulea*. *PLoS One* **8**(4): e62146. Doi: 10.1371/journal.pone.0062146.

Power, T., K. L. Clark, A. Harfenist, and D. B. Peakall, 1989. A review and evaluation of the amphibian toxicological literature. Technical Report 61. Hull, Quebec, Canada, Canadian Wildlife Service.

Pratihar, S. and J. K. Kundu, 2010. Hematological and immunological mechanisms of adaptation to hibernation in common Indian toad *Duttaphrynus melanostictus*. *Russian Journal of Herpetology* **17**: 97–100.

Quimby, F. W., A. C. Casey, and M. F. Arquette, 2005. From dogs to frogs: How pets, laboratory animals, and wildlife aided in elucidating harmful effects arising from a hazardous dumpsite. *ILAR Journal* **46**(4): 364–369.

Raffel, T. R., J. R. Rohr, J. M. Kiesecker, and P. J. Hudson, 2006. Negative effects of changing temperature on amphibian immunity under field conditions. *Functional Ecology* **20**(5): 819–828.

Raffel, T. R., J. M. Romansic, N. T. Halstead, T. A. McMahon, M. D. Venesky, and J. R. Rohr, 2013. Disease and thermal acclimation in a more variable and unpredictable climate. *Nature Climate Change* **3**(2): 146–151.

Reeder, A., G. Foley, D. Nichols, L. Hansen, B. Wikoff, S. Faeh, J. Eisold et al., 1998. Forms and prevalence of intersexuality and effects of environmental contaminants on sexuality in cricket frogs (*Acris crepitans*). *Environmental Health Perspectives* **106**(5): 261–266.

Regnault, C., I. Worms, C. Oger-Desfeux, C. Lima, S. Veyrenc, M. Bayle, B. Combourieu et al., 2014. Impaired liver function in *Xenopus tropicalis* exposed to benzo[a]pyrene: Transcriptomic and metabolic evidence. *BMC Genomics* **15**: 666. Doi: 10.1186/1471-2164-15-666.

Rhind, S. M., 2009. Anthropogenic pollutants: A threat to ecosystem sustainability? *Philosophical Transactions of the Royal Society of London. Series B: Biological Sciences* **364**(1534): 3391–3401.

Ribeiro, P. L., A. Camacho, and C. A. Navas, 2012. Considerations for assessing maximum critical temperatures in small ectothermic animals: Insights from leaf-cutting ants. *PLoS One* **7**(2) WOS:000302916800023.

Richter-Boix, A., G. A. Llorente, and A. Montori, 2006. Breeding phenology of an amphibian community in a Mediterranean area. *Amphibia-Reptilia* **27**(4): 549–559.

Rocha, P. L. and L. G. S. Branco, 1998. Physiological significance of behavioral hypothermia in hypoglycemic frogs (*Rana catesbeiana*). *Comparative Biochemistry and Physiology* **119A**(4): 957–961.

Rogers, K. D., M. B. Thompson, and F. Seebacher, 2007. Beneficial acclimation: Sex specific thermal acclimation of metabolic capacity in the striped marsh frog (*Limnodynastes peronii*). *Journal of Experimental Biology* **210**(16): 2932–2938.

Rohr, J. and K. Mccoy, 2010. A qualitative meta-analysis reveals consistent effects of atrazine on freshwater fish and amphibians. *Environmental Health Perspectives* **118**(1): 20–32.

Rohr, J. R., A. M. Schotthoefer, T. R. Raffel, H. J. Carrick, N. Halstead, J. T. Hoverman, C. M. Johnson et al., 2008. Agrochemicals increase trematode infections in a declining amphibian species. *Nature* **455**(7217): 1235-U1250.

Rollins-Smith, L. A., 2009. The role of amphibian antimicrobial peptides in protection of amphibians from pathogens linked to global amphibian declines. *Biochimica et Biophysica Acta—Biomembranes* **1788**(8): 1593–1599.

Saenz, D., L. A. Fitzgerald, K. A. Baum, and R. N. Conner, 2006. Abiotic correlates of anuran calling phenology: The importance of rain, temperature, and season. *Herpetological Monographs* (20): 64–82.

Santos, L. R. D., L. Franco-Belussi, and C. de Oliveira, 2011. Germ cell dynamics during the annual reproductive cycle of *Dendropophus minutus* (Anura: Hylidae). *Zoological Science* **28**(11): 840–844.

Sanuy, D. and P. Joly, 2009. Olfactory cues and breeding habitat selection in the natterjack toad, *Bufo calamita*. *Amphibia-Reptilia* **30**(4): 555–559.

Sapolsky, R. M., 2002. Endocrinology of the stress-response. In Becker, J. B., S. M. Breedlove, D. Crews, and M. M. McCarthy (eds), *Behavioral Endocrinology*. London, MIT: 409–450.

Sapolsky, R. M., L. M. Romero, and A. U. Munck, 2000. How do glucocorticoids influence stress responses? Integrating permissive, suppressive, stimulatory, and preparative actions. *Endocrine Reviews* **21**(1): 55–89.

Schiesari, L., B. Grillitsch, and H. Grillitsch, 2007. Biogeographic biases in research and their consequences for linking amphibian declines to pollution. *Conservation Biology* **21**(2): 465–471.

Schmidt-Nielsen, K., 1972. *How Animals Work*. New York, Press Syndicate of the University of Cambridge.

Schulz, R., 2004. Field studies on exposure, effects, and risk mitigation of aquatic nonpoint-source insecticide pollution: A review. *Journal of Environmental Quality* **33**(2): 419–448.

Seebacher, F., J. A. Tallis, and R. S. James, 2014. The cost of muscle power production: Muscle oxygen consumption per unit work increases at low temperatures in *Xenopus laevis*. *Journal of Experimental Biology* **217**(11): 1940–1945.

Selcer, K. W. and J. D. Verbanic, 2014. Vitellogenin of the northern leopard frog (*Rana pipiens*): Development of an ELISA assay and evaluation of induction after immersion in xenobiotic estrogens. *Chemosphere* **112**: 348–354.

Sifakis, S. and A. Tsatsakis, 2011. Pesticide exposure and health related issues in male and female reproductive system. Biomarkers of exposure and effects on pregnancy. *Toxicology Letters* **205**: S5–S5.

Sillero, N. and P. Tarroso, 2010. Free GIS for herpetologists: Free data sources on Internet and comparison analysis of proprietary and free/open source software. *Acta Herpetologica* **5**(1): 63–85.

Simmaco, M., A. Boman, M. L. Mangoni, G. Mignogna, R. Miele, D. Barra, and H. G. Boman, 1997. Effect of glucocorticoids on the synthesis of antimicrobial peptides in amphibian skin. *FEBS Letters* **416**(3): 273–275.

Sinervo, B., 2010. Erosion of lizard diversity by climate change and altered thermal niches. *Science* **328**(5984): 894–899.

Singha, U., N. Pandey, F. Boro, S. Giri, A. Giri, and S. Biswas, 2014. Sodium arsenite induced changes in survival, growth, metamorphosis and genotoxicity in the Indian cricket frog (*Rana limnocharis*). *Chemosphere* **112**: 333–339.

Sinsch, U., 2006. Orientation and navigation in amphibia. *Marine and Freshwater Behaviour and Physiology* **39**(1): 65–71.

Sinsch, U., N. Oromi, and D. Sanuy, 2007. Growth marks in natterjack toad (*Bufo calamita*) bones: Histological correlates of hibernation and aestivation periods. *Herpetological Journal* **17**(2): 129–137.

Sodhi, N. S., D. Bickford, A. C. Diesmos, T. M. Lee, L. P. Koh, B. W. Brook, C. H. Sekercioglu, and C. J. A. Bradshaw, 2008. Measuring the meltdown: Drivers of global amphibian extinction and decline. *PLoS One* **3**(2): e1636. DOI:1610.1371/journal.pone.0001636.

Solomon, K., J. Carr, L. Du Preez, J. Giesy, R. Kendall, E. Smith, and G. Van Der Kraak, 2008. Effects of atrazine on fish, amphibians, and aquatic reptiles: A critical review. *Critical Reviews in Toxicology* **38**(9): 721–772.

Somero, G. N., 2011. Comparative physiology: A "crystal ball" for predicting consequences of global change. *American Journal of Physiology—Regulatory Integrative and Comparative Physiology* **301**(1): R1–R14.

Storey, K. B. and J. M. Storey, 1996. Natural freezing survival in animals. *Annual Review of Ecology and Systematics* **27**: 365–386.

Stuart, S. N., J. S. Chanson, N. A. Cox, B. E. Young, A. S. L. Rodrigues, D. L. Fischman, and R. W. Waller, 2004. Status and trends of amphibian declines and extinctions worldwide. *Science* **306**(5702): 1783–1786.

Swanson, D. L. and S. L. Burdick, 2010. Overwintering physiology and hibernacula microclimates of Blanchard's cricket frogs at their Northwestern range boundary. *Copeia* (2): 247–253.

Swanson, D. L., B. M. Graves, and K. L. Koster, 1996. Freezing tolerance/intolerance and cryoprotectant synthesis in terrestrially overwintering anurans in the Great Plains, USA. *Journal of Comparative Physiology B* **166**(2): 110–119.

Szuroczki, D. and J. M. L. Richardson, 2009. The role of trematode parasites in larval anuran communities: An aquatic ecologist's guide to the major players. *Oecologia* **161**(2): 371–385.

Tattersall, G. J. and G. R. Ultsch, 2008. Physiological ecology of aquatic overwintering in ranid frogs. *Biological Reviews* **83**(2): 119–140.

Tavera-Mendoza, L., S. Ruby, P. Brousseau, M. Fournier, D. Cyr, and D. Marcogliese, 2002. Response of the amphibian tadpole *Xenopus laevis* to atrazine during sexual differentiation of the ovary. *Environmental Toxicology and Chemistry* **21**(6): 1264–1267.

Taylor, B., D. Skelly, L. Demarchis, M. Slade, D. Galusha, and P. Rabinowitz, 2005. Proximity to pollution sources and risk of amphibian limb malformation. *Environmental Health Perspectives* **113**(11): 1497–1501.

Titon, B., Jr., C. A. Navas, J. Jim, and F. R. Gomes. 2010. Water balance and locomotor performance in three species of neotropical toads that differ in geographical distribution. *Comparative Biochemistry and Physiology* **156A**: 129–135.

Todd, B. D., D. E. Scott, J. H. K. Pechmann, and J. W. Gibbons, 2011. Climate change correlates with rapid delays and advancements in reproductive timing in an amphibian community. *Proceedings of the Royal Society B—Biological Sciences* **278**(1715): 2191–2197.

Tournefier, A., 1982. Corticosteroid action on lymphocyte sub-populations and humoral immune response of axolotl (Urodele Amphibian). *Federation Proceedings* **41**(3): 608.

Trumbo, D. R., A. A. Burgett, and J. H. Knouft, 2011. Testing climate-based species distribution models with recent field surveys of pond-breeding amphibians in eastern Missouri. *Canadian Journal of Zoology—Revue Canadienne De Zoologie* **89**(11): 1074–1083.

Tsai, P. S., J. B. Lunden, and J. T. Jones, 2003. Effects of steroid hormones on spermatogenesis and GnRH release in male Leopard frogs, Rana pipiens. *General and Comparative Endocrinology* **134**: 330–338.

Van Straalen, N. M., 2003. Ecotoxicology becomes stress ecology. *Environmental Science & Technology* **37**(17): 324A–330A.

Venesky, M. D., T. E. Wilcoxen, M. A. Rensel, L. Rollins-Smith, J. L. Kerby, and M. J. Parris, 2012. Dietary protein restriction impairs growth, immunity, and disease resistance in southern leopard frog tadpoles. *Oecologia* **169**(1): 23–31.

Venturino, A., E. Rosenbaum, A. C. De Castro, O. L. Anguiano, L. Gauna, T. F. De Schroeder, and A. M. P. De D'Angelo, 2003. Biomarkers of effect in toads and frogs. *Biomarkers* **8**(3–4): 167–186.

Visser, M. E., 2008. Keeping up with a warming world; assessing the rate of adaptation to climate change. *Proceedings of the Royal Society B—Biological Sciences* **275**(1635): 649–659.

Vittoz, P., D. Cherix, Y. Gonseth, V. Lubini, R. Maggini, N. Zbinden, and S. Zumbach, 2013. Climate change impacts on biodiversity in Switzerland: A review. *Journal for Nature Conservation* **21**(3): 154–162.

Wake, D. B. and V. T. Vredenburg, 2008. Are we in the midst of the sixth mass extinction? A view from the world of amphibians. *Proceedings of the National Academy of Sciences of the United States of America* **105**: 11466–11473.

Ward, C. K., C. Fontes, C. W. Breuner, and M. T. Mendonca, 2007. Characterization and quantification of corticosteroid-binding globulin in a southern toad, *Bufo terrestris*, exposed to coal-combustion-waste. *General and Comparative Endocrinology* **152**(1): 82–88.

Warne, R. W., E. J. Crespi, and J. L. Brunner, 2011. Escape from the pond: Stress and developmental responses to ranavirus infection in wood frog tadpoles. *Functional Ecology* **25**(1): 139–146.

Wells, K. D., 2001. The energetics of calling in frogs. In Ryan M. J. (ed.), *Anuran Communication*. Washington, DC, Smithsonian Institution: 45–60.

Wells, K. D. and C. Bevier, 1997. Contrasting patterns of energy substrate use in two species of frogs that breed in cold weather. *Herpetologica* **53**: 70–80.

Wells, K. D. and T. L. Taigen, 1989. Calling energetics of a neotropical treefrog, *Hyla microcephala*. *Behavioral Ecology and Sociobiology* **25**: 13–22.

Wharton, D. A., 2011. Cold tolerance of New Zealand alpine insects. *Journal of Insect Physiology* **57**(8): 1090–1095.

Wikelski, M. and S. J. Cooke, 2006. Conservation physiology. *Trends in Ecology & Evolution* **21**(1): 38–46.

Wilczynski, W. and K. S. Lynch, 2011. Female sexual arousal in amphibians. *Hormones and Behavior* **59**(5): 630–636.

Wilson, R. S. and C. E. Franklin, 1999. Thermal acclimation of locomotor performance in tadpoles of the frog *Limnodynastes peronii*. *Journal of Comparative Physiology B—Biochemical Systemic and Environmental Physiology* **169**: 445–451.

Wingfield, J. C. and R. M. Sapolsky, 2003. Reproduction and resistance to stress: When and how. *Journal of Neuroendocrinology* **15**(8): 711–724.

Woods, H. A., M. F. Poteet, P. D. Hitchings, R. A. Brain, and B. W. Brooks, 2010. Conservation physiology of the Plethodontid salamanders *Eurycea nana* and *E. sosorum*: Response to declining dissolved oxygen. *Copeia*(4): 540–553.

Wright, R. K. and E. L. Cooper, 1981. Temperature effects on ectotherm immune responses. *Developmental and Comparative Immunology* **5**(Suppl. 1): 117–122.

Yadav, S., S. Giri, U. Singha, F. Boro, and A. Giri, 2013. Toxic and genotoxic effects of roundup on tadpoles of the Indian skittering frog (*Euflictis cyanophlyctis*) in the presence and absence of predator stress. *Aquatic Toxicology* **132**: 1–8.

Yaron, Z. and B. Siva, 2006. Reproduction. In Evans D. H. and J. B. Claiborne (eds), *The Physiology of Fishes*. Boca Raton, FL, CRC Press/Taylor & Francis: 343–386.

Zapata, A. G., A. Varas, and M. Torroba, 1992. Seasonal variations in the immune system of lower vertebrates. *Immunology Today* **13**(4): 142–147.

Zhelev, Z. M., G. S. Popgeorgiev, and M. V. Angelov, 2013. Investigating the changes in the morphological content of the blood of *Pelophylax ridibundus* (Amphibia: Ranidae) as a result of anthropogenic pollution and its use as an environmental bioindicator. *Acta Zoologica Bulgarica* **65**(2): 187–196.

Zimmerman, S. L., J. Frisbie, D. L. Goldstein, J. West, K. Rivera, and C. M. Krane, 2007. Excretion and conservation of glycerol, and expression of aquaporins and glyceroporins, during cold acclimation in Cope's gray tree frog Hyla chrysoscelis. *American Journal of Physiology—Regulatory Integrative and Comparative Physiology* **292**(1): R544–R555.

chapter eight

Assessing the physiological sensitivity of amphibians to extreme environmental change using the stress endocrine responses

Edward J. Narayan

Contents

Introduction

Amphibians are excellent bioindicators of environmental health, and being ectothermic animals, their day–day life-history traits are influenced by both short- and long-term changes in environmental temperature (Beebee and Griffiths 2005). Unfortunately, the population of various amphibian species have started to decline around the world and many species have already become extinct or have been placed on the International Union for Conservation of Nature (IUCN) endangered species list. Since 1980, declines in certain amphibian populations have occurred worldwide (Crump et al. 1992). In 2004, amphibian biologists at an international conference announced that 32% of amphibian species are currently threatened, the population of 44% of species are in a decline, and 120 amphibian species have likely become extinct only in the last 25 years (Stuart et al. 2004). Factors such as habitat loss and deterioration, global warming, ultraviolet light, acid rain, commercial collection, invasive species, and pesticide use have all been investigated and implicated as causes for these declines in amphibian populations. Global amphibian populations are declining rapidly toward extinction and scientists have begun analysis of past declines to make predictions about future declines based on complex climatic models (Bosch et al. 2007; Raffel et al. 2012), with predictions made that anthropogenic-induced climate change continues to have huge impacts on amphibian populations worldwide. Furthermore, the additional impacts of pathogenic diseases and other human-induced factors, such as habitat destruction and urbanization are making amphibian population declines even more catastrophic. Global mean temperatures have risen by approximately 0.74°C over the past 100 years with the Fourth Intergovernmental Panel on Climate Change (IPCC) reports predicting that without intervention, this trend will continue (Root et al. 2003; Parmesan 2006; IPCC 2007a). By the end of this century, global temperatures are predicted to increase by 1.8–4°C, with higher latitudes having the greatest warming (IPCC 2007b). It is predicted that climate change will cause major environmental and economic impacts, particularly from increases in the frequency of extreme weather events such as bushfire, droughts, floods, and heat waves, for example, in Australia (Hughes 2003; IPCC 2007a,b). It is plausible that greater frequency of extreme changes in global climate within the last 50 years has led to rapid amphibian declines (Alexander and Eischeid 2001; Beebee and Griffiths 2005). Possible evidence of this decline is validated by the golden toad (*Incilius periglenes*), which has not been seen in the Costa Rican rainforest since the early 1980s (Pounds et al. 1997; Beebee and Griffiths 2005). *I. periglenes* was found in the higher altitudes of the mountains, and species restricted to lower altitudes have since then moved higher into the mountains (Wake and Vredenburg 2008). A major concern associated with global climate change is that with a relatively large number of frogs and lizards identified as high-elevation specialists, species restricted to higher altitudes have nowhere to go in response to extreme variation in environmental temperature. Even for lowland species, possible areas of dispersal in response to anthropogenic-induced climate change are constantly affected by the processes associated with habitat loss and fragmentation (Pearson and Dawson 2005; Nyström et al. 2007; Laurance et al. 2011). However, climate variability could also contribute to infectious diseases, such as chytridiomycosis, which is contributing toward mass mortalities of amphibian species globally. The chytrid fungus (*Batrachochytrium dendrobatidis*) affects postmetamorphic amphibians (newly metamorphosed individuals are at the highest risk of lethal chytrid fungus infection) caus-ing disruptions to important bodily functions, for example, osmoregula-tion and respiration (Bozinovic et al. 2011). Amphibians can potentially be at higher risk of infection following a weakened immune response as well as increased susceptibility to the pathogen due to chronic stress (Carey et al. 1999; Rollins-Smith 2001; Beebee and Griffiths 2005; Narayan and Hero 2014; Bovo et al. 2016).

As our planet gets warmer, the so-called "sixth mass extinction" will worsen, putting immense pressure on living organisms to adjust their intrinsic physiology to promote their survival (Wake and Vredenburg 2008). Amphibians are more susceptible to extreme thermal environments than other species because of their permeable skin, aquatic–terrestrial life cycle, and membranous unshelled eggs (Carey and Alexander 2003). Impacts of anthropogenic-induced environmental change in amphib-ian species can be categorized as direct or indirect as follows: breeding phenology of amphibians are directly affected by changes in climate. For example, it was hypothesized that amphibians affected by global warm-ing tend to show early breeding as the average temperatures increase. This could increase species vulnerability toward loss of young due to earlier snowmelt-induced floods and early season freezes (Beebee 1995; Blaustein et al. 2001; Gibbs and Breisch 2001). Furthermore, there is strong evidence that climatic irregularities have contributed to the massive col-lapse of amphibian populations around the world. For example, studies from eastern Australia (Ingram 1990; Laurance et al. 1996) found correla-tions between drought and massive declines of stream-dwelling rainfor-est amphibians. Corn and Fogleman (1984) found a correlation between the extinction of the mountain population of the northern leopard frog, *Rana pipiens*, and drought in North America. Severe drought has been the reason for dramatic population declines of the Puerto Rican coqui frog, the *Eleutherodactulus coqui* (Stewart 1995). Indirect effects of climate change include influences on breeding cycles, diseases, and combination with other environmental factors (e.g., urbanization) (Carey and Alexander 2003). As highlighted, climate change is impacting amphibian populations in complex ways through synergistic interactions with other extreme envi-ronmental factors such as infectious diseases. For instance, local changes in the environment can decrease immune function and lead to pathogen outbreaks and elevated mortality. Gervasi and Foufopoulos (2008) demon-strated that the amphibian immune system becomes compromised due to a faster rate of metamorphosis under increased risk of pond desiccation to favor survival. Amphibian immune system function, including innate and adaptive immune responses, depends on the external temperature, although amphibians may be unable to quickly adjust these physiologi-cal systems under extreme thermal environments, leading to suboptimal immune functions (Maniero and Carey 1997; Raffel et al. 2006). The colder aspect of extreme thermal variation is also important because seasonal reductions in environmental temperature tend to reduce amphibian behaviors and also cause reductions in peripheral leukocyte levels, lymphocyte (T and B cells) proliferation, macrophage endocytosis, and abundance of antimicrobial skin peptides, which is a vital immune response against chytrid fungus infection (Bly and Clem 1991; Pxytycz and Jozkowicz 1994; Maniero and Carey 1997; Matutte et al. 2000; Raffel et al. 2006). Decreases in environmental temperature might be increasing the severity of pathogenic infections in amphibian species, especially by increasing the rates of release of chytrid fungus zoospores (Woodhams et al. 2008). Furthermore, global warming is supporting the prevalence of pathogenic diseases, such as chytridiomycosis, because of warmer nighttime temperatures and mistier daytime conditions, resulting in a better growth spectrum for chytrid fungus (Pounds et al. 1999). It is possible that weakened or disrupted physiological processes, such as immune system and metabolic functions, could be a "down-stream" consequence of chronic stress. It is widely accepted that stress hormones and the neuroendocrine stress axis are principal mediators of physiological and behavioral responses to environmental change in animals (Denver 2009).

Glucocorticoids are steroid hormones, which are highly conserved across vertebrate groups and released to enable animals to cope with environmental challenges. Therefore, baseline and stress-induced glucocorticoids change with respect to varying environmental factors (Narayan et al. 2010; Ouyang et al. 2011). The stress endocrine axis in vertebrates is tightly linked with the immune system, and stress hormones also regulate metabolic processes in amphibians (Wack et al. 2012; de Assis et al. 2015). Chronic elevation of glucocorticoids could eventually minimize the coping capacity of amphibians against novel stressors or more virulent strains of pathogenic infections (Phillips and Puschendorf 2013; Narayan and Hero 2014). Therefore, quantification of the stress endocrine responses of amphibians during exposure to thermal stress under controlled laboratory conditions and/or field conditions can provide new knowledge on "sublethal" impacts of extreme environmental change on amphibian species. Given that extinction rates for amphibians are faster than for any other vertebrate group, there is an urgent need to advance our understanding of amphibian stress physiology and align this knowledge with conservation efforts.

The primary aim of this chapter is to update the information on the physiological stress responses of amphibian species to variations in thermal environments, with a focus on their endocrine response. Second, this review describes a new quantitative measure of the plasticity of stress endocrine responses to acute shifts in environmental temperature toward extremes, using the concept of "glucocorticoid thermal reaction norms." The importance of using noninvasive hormone monitoring tools to study the variation in the physiological stress responses of amphibian species with respect to environmental change is also examined.

Stress physiology

Physiological stress responses to environmental stressors are highly conserved across vertebrate groups and involve an endocrine cascade resulting in the production of glucocorticoids. In 1936, Hans Selye characterized the stress response as a common set of generalized physiological responses that were experienced by all organisms exposed to a variety of environmental challenges, such as thermal changes or exposure to noise. Decades later, Sterling and Eyer (1988) introduced the term "allostasis," which has been heavily advocated by McEwen and Stellar (1993) as the body's ability to adapt to a changing environment in situations that did not challenge survival. Baseline glucocorticoids or baseline stress hormones are defined as the "level of hormone secretion associated with normal physiological function." At baseline levels, glucocorticoids are secreted to maintain allostasis (homeostasis) in response to predictable energetic demands, resulting primarily in the binding of type I glucocorticoid receptors (Romero 2004). Stress-induced glucocorticoid levels or physiological stress responses are defined as "increases in glucocorticoid levels beyond baseline levels," resulting in saturation of the low-capacity, high-affinity type I glucocorticoid receptors and binding of the high-capacity, low-affinity type II receptors. According to Romero (2004), baseline and stress-induced concentrations of glucocorticoids have completely different physiological and behavioral effects and can even be thought of as reflecting two complementary hormonal systems. Furthermore, under the influence of extreme environmental change, glucocorticoid responses do not change over time or become chronically elevated or lead to chronically diminished functioning of the entire stress endocrine system.

The differences between baseline levels and physiological stress responses have been explained in three physiological states: allostasis, allostatic load, and allostatic overload (McEwen and Wingfield, 2007). Allostasis refers to achieving stability through change and baseline stress hormone levels support allostasis via the regulation of energy distribution to vital systems such as predictable environmental changes and life-history stages. Allostatic load refers to the cumulative result of maintaining allostasis and accounts for the seasonal increases in stress hormones caused by predictable increases in energy requirements during more demanding periods, such as winter and breeding. Allostatic overload occurs when an unpredictable event increases energy demands beyond a point at which energy modulation can successfully deal with it. While an acute stress response is necessary to ensure survival and adjustment to changes in the environment, it is only a concern when the physiological stress response becomes chronic and threatens animal well-being by exerting deleterious effects on the individual's biological state. This has welfare implications during energetically costly periods, such as growth and reproduction, or when animals are suffering from parasitic infections or other diseases (Monfort 2003; Kindermann et al. 2012). In the context of this review chapter, I will refer to the integrated stress response, which gives the complete measure of the stress endocrine response during exposure to a physical or psychological stressor in an animal, removing any effects of variation in baseline stress hormone levels. The integrated stress responses have been used in the context of chronic stress in wildlife (Dickens et al. 2009).

In amphibians, the physiological stress response involves activation of the hypothalamo–pituitary–interrenal (HPI) axis culminating with the secretion of glucocorticoids in response to any kind of stressor. The process begins from the hypothalamus in which the corticotrophin releasing factor (CRF) acts on the anterior pituitary gland to release adrenocorticotropic hormone (ACTH) into the peripheral circulation. ACTH in turn acts on the interrenal glands and stimulates it to synthesize and secrete corticosterone which are involved in the modulation of energy mobilization from the interrenal glands. The HPI-axis responses are consistent both across taxonomic groups (Chambers et al. 2013) and across developmental stages (Glennemeier and Denver 2002). Corticosterone secretions have direct implications for physiological and behavioral responses to different types of stimuli in amphibians (Narayan et al. 2010). Stress hormone responses can give way to physiological and behavioral adjustments, which may vary between minimum and maximum responses to different stressors, that allow an individual to cope with stress acute and chronic responses. For example, handling stress leading to acute corticosterone elevations is an immediate response to a stressor and not a chronic elevation (Narayan et al. 2012). Physiological stress responses can result in changes in behavior, reproduction, reproductive hormones, metabolism, immune responses, and nutrition. Furthermore, when the physiological stress response occurs, less important functions at that time, such as foraging behavior, may be temporarily suppressed (Bliley and Woodley 2012). On the other hand, Bliley and Woodley (2012) found that repeated exposure to handling stressors and elevation of corticosterone concentrations in the plasma of a salamander (*Desmognathus ocoee*) still resulted in most individuals courting and mating. This may be a reflection of the species, breeding behaviors being energetically inexpensive, but indicates that organisms under stress are still able to express normal breeding behaviors (Bliley and Woodley 2012).

Measuring stress hormones

The use of blood has been used widely for assessing amphibian endocrine function. In 1961, Carstensen and colleagues demonstrated *in vitro* that corticosterone and aldosterone were the primary steroids formed in incubates of the interrenal glands of the American bullfrog (*Rana catesbeiana*) by stimulating their production using mammalian ACTH (Carstensen et al. 1961). Dale (1962) most likely had conducted the earliest work on glucocorticoid measurement in anuran excreta through his study on the leopard frog, *R. pipiens* (Schreber), and the northern wood frog, *Rana sylvatica* (Le Conte), larvae. In another study, Jungreis et al. (1970) identified corticosterone and aldosterone (3:1 ratio) as major hormone products of the interrenal tissues of *R. pipiens*. Interestingly, earlier studies have also reported that fully aquatic amphibians produce cortisol as their primary glucocorticoid, while terrestrial or semiterrestrial amphibians produce corticosterone as their primary glucocorticoid (Bern and Nandi 1964; Nandi 1967). In their study, Jungreis et al. (1970) suggested a potential ontogenetic shift in the steroidogenic pathway from cortisol in the aquatic larvae to corticosterone in the semiterrestrial or terrestrial adult. A few years later, Leboulenger et al. (1977) measured corticosterone in plasma samples of the edible frog (*Rana esculenta*) via a competitive protein-binding radioimmunoassay (RIA) method using baboon plasma as the corticosterone binding globulin source. Their study provided the first demonstration of the seasonal pattern of plasma corticosterone in anurans. Two years later, Dupont and colleagues provided the first ever evidence of circadian and circannual variation in corticosterone in mature male and female free-living *R. esculenta* (Dupont et al. 1979). Subsequently, Leboulenger et al. developed two RIA techniques for the direct measurement of frog plasma concentrations of corticosterone (Leboulenger et al. 1982). Several studies used the RIA technique to measure plasma corticosterone in anurans. For example, Pancak and Taylor (1983) studied the daily and seasonal rhythms of plasma corticosterone in American toads (*Bufo americanus*). Their study also provided insights into the role of corticosterone in amphibian locomotion. Later, Licht et al. (1983) demonstrated the effects of capture and captivity on plasma levels of gonadal steroids and on corticosterone in the bullfrog (*R. catesbeiana*). Enzyme immunoassays (EIAs) have replaced traditional methods such as RIA as they are cost effective. Direct-ELISA, also known as single-antibody EIA, is as sensitive as RIA and is gaining popularity for the measurement of stress hormone metabolites in amphibians (Narayan et al. 2010; Janin et al. 2012; Kindermann et al. 2012).

Noninvasive hormone monitoring

Noninvasive monitoring of hormones refers to the technique of quantification of hormonal metabolites in animals, using biological samples (end-products such as urine, feces, saliva, hair, fur, etc.) that were collected using passive sampling (i.e., without causing undue stress to the animal as would normally occur during more invasive methods, such as cardiac puncture for serum sampling). For anurans, urine sampling provides a simple method of assessing baseline and short-term corticosterone stress responses in anurans with minimal disturbance. Mild capture and handling (for a maximum of 5 min on each sampling occasion) causes increased urinary corticosterone metabolite concentrations in anurans (Narayan et al. 2010), which enables individual baseline and short-term corticosterone stress responses to be assessed over time periods (Narayan et al. 2011). This type of data is needed to better understand the role of corticosterone in relation to the interaction of amphibians and their environment and potential consequences to fitness. Furthermore, corticosterone metabolites can also be measured using another noninvasive technique in amphibian species, such as saliva and aquatic media (Janin et al. 2011; Gabor et al. 2013).

During handling of amphibians for urine collection, they should always be manupulated by wearing new nonpowdered gloves or disposable freezer bags for each individual to avoid the spread of chytrid fungus. Urine can be obtained upon initial capture and by gentle massage of the underbelly abdomen over a sterile cup. This method works well for large anurans (such as cane toads, *Rhinella marina*; snout–vent length = 100 mm). Another simple method, using sterile pipette tips (200 μL) works reasonably well for large- to medium-sized species, such as the stony creek frog (*Litoria wilcoxii*; snout–vent length = 45–70 mm), red-eyed tree frog (*Litoria chloris*; snout–vent length = 65 mm), and for the great barred frog (*Mixophyes fasciolatus*; snout–vent length = 80 mm). Typically, the sterile tip of the pipette is inserted at a very gentle pace inside (5 mm) the anuran's cloaca while the frog is manually held (this can be done by the same person manually holding the anuran), and urine is obtained by capillary action. For medium- to small-sized species, such the *Philoria* species, a nonheparinized microcapillary tube (20–50 μL) could be used. Sigma–Aldrich (#P2174) microcapillary tubes can be used for the purpose of collecting urine in small frog species, such as *Pseudophryne* and *Crinia* species. The time required for urine collection from initial capture to sampling is normally between 30 s and 5 min.

Physiological validation

It is crucial to demonstrate that noninvasive stress hormone measures accurately reflect biological events of interest. Administration of ACTH mimics a natural interrenal stress response by causing a rapid rise in circulating corticosterone, followed by a return to baseline within a few hours. The same pattern should also occur in urine or feces, with the onset of peak excretion delayed by the species-specific excretory lag time (Wasser et al. 2000; Preest et al. 2005). This area of investigation is pretty much lacking for amphibians, however, as highlighted biological/physiological validation is important when attempting to develop a new noninvasive steroid metabolite assay for any animal species. Biological validation determines the excretory lag time between stimulation of an endocrine gland and the appearance of its hormone metabolites in excreta. Presently, urinary corticosterone metabolite EIAs have been validated physiologically for three anuran species, the stony creek frog (*L. wilcoxii*), great barred frog (*M. fasciolatus*), and the Fijian ground frog (*P. vitiana*) (Narayan et al. 2010; Kindermann et al. 2012; Graham et al. 2013). These studies demonstrated that the application of an ACTH challenge protocol induces a raise in urinary corticosterone followed by its return to baseline levels within 2 to 3 days (Narayan et al. 2010; Kindermann et al. 2012; Graham et al. 2013). The relatively slow (several hours to a few days) lag time of urinary corticosterone metabolites means that it can be used to measure the physiological stress responses to both short-term and chronic stressors. Future studies could also attempt to biologically validate their assays using alternative means if an ACTH challenge is impossible (e.g., when working with a critically endangered species). For example, a moderate handling stressor as opposed to mild handling stressor has been shown to produce significantly different levels of glucocorticoids in anurans (Narayan et al. 2012).

Thermal influence on the stress endocrine responses of amphibians

The intrinsic reproductive and stress endocrine systems of amphibians work synchronously with environmental temperatures both daily and seasonally, especially during the breeding period (Lofts et al. 1972; Licht et al. 1983). The simplest way to measure amphibian body temperature is with noncontact temperature measurement using a portable handheld infrared thermometer. The sensitivity of the HPI axis to environmental temperature was demonstrated clearly in the study by Jurani et al. (1973). The study showed that frogs exposed to 30°C and 40°C had a significant increase in plasma corticosterone. At acclimation temperature of 1°C, ACTH injection of frogs did not increase plasma corticosterone either 30 or 90 min after administration. However, significant increases in plasma corticosterone were seen at 30 min after ACTH administration when frogs were acclimated at 10°C. Furthermore, the increase in plasma corticosterone was even more significant after 30 min in frogs kept at 20°C or 30°C, but no further increase in plasma corticosterone occurred at temperatures up to 40°C. Reading (2007) found that female body condition and survivorship of palmate newts (*Lissotriton helveticus*) were negatively impacted due to the effects of extreme temperatures and warmer climates leading to decreased body condition. Galloy and Denoël (2010) showed that increases in water temperature can negatively influence the fecundity of palmate newts.

In a recent study, it was demonstrated using the cane toad (*R. marina*) that acclimation at different temperatures (15°C, 25°C, or 35°C) resulted in significant differences in baseline and short-term corticosterone responses (assessed using standard capture handling protocol). Baseline and stress-induced corticosterone levels were lower in toads acclimated to 15°C compared to those acclimated to 35°C (Narayan et al. 2012).

The most likely explanation for the thermodynamic effects on the endocrine system and the stress response follows that synthesis of corticosterone in the interrenal gland requires supply of the adrenal cortical cells with ACTH through blood circulation and exposure to high environmental temperature causing increased blood flow to the interrenal glands that increases the rate of release of corticosteroids (Hume and Egdahl 1959). It has been shown that acute increase in ambient temperature from 5°C to 30°C led to sixfold increases in corticosterone production by the interrenal glands of the marsh frog (*Rana ridibunda*) (Leboulenger et al. 1978). Recently, Narayan and Hero (2014) showed that exposure to acute thermal stress by acclimating cane toads to mildly warm temperature (25°C) and then to acute temperature treatments of 30°C, 35°C, or 40°C (hypothetical acute thermal stressors) led to increases in urinary corticosterone metabolites.

More studies are warranted, using more threatened species to determine whether similar thermodynamic effects, as seen in the introduced cane toads, are present in other species of amphibians. Furthermore, there are species, such as cane toads, adapted to living in extreme arid environments, that are capable of living in a climatically extreme environment with prolonged periods of dry weather and extreme heat (Florance et al. 2011; Tingley et al. 2012). It was shown that cane toads living at the migration front in the Australian desert tend to minimize their acute physiological stress responses during the extremely hot summer months, enabling them to miminize cutaneous water loss (as corticosterone is known as an osmoregulatory hormone) (Jessop et al. 2013).

In nature, the thermodynamic effects on the HPI axis can be studied using altitudinal gradients as surrogate for climatic variability (Bears et al. 2003). Interestingly, it has been demonstrated in field studies that amphibians, such as the great barred frog (*M. fasciolatus*), living along altitudinal gradients (highlands) in Southeast Queensland, Australia, expressed significantly higher levels of baseline urinary corticosterone metabolites compared to their lowland counterparts (Graham et al. 2013). Physiological endocrine measures can be used in combination with ecological data (e.g., chytrid fungus prevalence) and environmental data to better understand the responses of amphibians to changes in environmental temperature and their susceptibility to pathogenic infections.

Thermal stress is a physical stressor and it has been shown in mammalian models that several brain nuclei are involved in the integration and triggering of the stress response to high thermal stress, with hypothalamic centers playing the pivotal role. Future studies should also examine the crosstalk between the stress endocrine response and other physiological endocrine systems (McNab 2002). For example, arginine-vasopressin, which is involved in the regulation of osmotic balance, is also known to influence the stress response during exposure to acute thermal stress in other animal models (Jasnic et al. 2013). As heat stress occurs, there will be multiple effects on the different axes of the endocrine systems, such as the hypothalamus–pituitary–adrenal (HPA) axis, the hypothalamus–pituitary–thyroid axis, or the hypothalamus–pituitary–gonadal axis. Among them, the main feature of the stress reaction is the activation of the HPA axis leading to the increase of ACTH and glucocorticoids (Jasnic et al. 2013). There would also be potential for "down-stream" effects on the immune response because ACTH regulates the production of cytokines.

Future studies based on the exposure to high thermal stress should focus on multiple physiological, endocrine, and biochemical processes that can be impacted by extreme environmental stress. These types of physiological manipulations can be done using phenotypic engineering (e.g., altering the rates of corticosterone synthesis and glucocorticoid (GC)-receptor activity) (Jessop et al. 2013).

Predicting amphibian physiological sensitivity to extreme environmental change

Chronic thermal stress as a result of global warming and extreme thermal events such as heat waves could affect the performance and fitness traits of ectotherms (Francisco et al. 2011; Nguyen et al. 2011; Coumou and Robinson 2013). Stress endocrine system monitoring could provide crucial insights into the possible sublethal effects of thermal stress on the physiology of amphibians. Corticosterone stress responses are essential for coping with acute stressors such as predatory encounters through changes in metabolic rates (Wack et al. 2012) and behavioral changes (Clinchy et al. 2013). They are also essential for combating pathogenic infections and injury through increased immunocompetence (Hopkins and DuRant 2011; Kindermann et al. 2012). Chronic or repeated exposure to high temperature can lead to attenuation or down-regulation of the corticosterone stress response as demonstrated under laboratory conditions using the cane toad model (Narayan and Hero 2014). Down-regulation is defined in the stress endocrinology literature as the reduction in the magnitude of acute stress hormone responses in animals with exposure to a repeated stressor (Rich and Romero 2005). This result suggests that exposure to unpredictable environmental temperatures could affect the physiology of amphibians through modulation of the stress endocrine axis.

Thus, chronic stress due to exposure to extreme environmental stressors could lead to down-stream impacts on reproduction and survival.

Glucocorticoid thermal reaction norms

Baseline and stress-induced glucocorticoid responses are evolutionary liable physiological traits; hence they have a great deal of intra- and inter-population variation, and this variation could provide a better under-standing of how animals might be coping with extreme environmental changes (Dingemanse et al. 2010). Selection of optimal physiological traits could be important for survival under the influence of extreme environmental conditions for amphibian species. The relationship between thermal environmental variation and glucocorticoids can be studied to provide better understanding of the thermal tolerance of amphibian species against shifts in thermal environments caused by extreme environmental changes. This relationship can be presented using a concept known as thermal reaction norms. Thermal reaction norms provide a quantitative measure of thermal tolerance in animals, using a wide range of functions from biochemical pathways to whole animals (Prosser 1973; Hochachka and Somero 2002). This concept can be applied under natural conditions, such as across latitudes and altitudes, and the various patterns of the thermal reaction norm curves can be used for comparing the thermal sensitivity of animal species and populations across environments (e.g., Huey and Kingsolver 1989; Morley et al. 2012). This concept can be applied to study the thermodynamic effects of thermal variation on the stress endocrine response, using "glucocorticoid thermal reaction norms." A glucocorticoid thermal reaction norm will define the strength of the relationship between environmental temperature and stress response titers (integrated responses). The magnitude of corticosterone stress response is measured using the area under the curve and referred to as the integrated corticosterone response. The total area under the curve minus the area attributed to corticosterone concentrations at baseline (0 min) is expressed as the corrected integrated corticosterone response. Integrated corticosterone response accounts for type II glucocorticoid receptors, which are bound only when the stress-induced glucocorticoid concentrations are high, and they regulate classic functions of glucocorticoids (Romero 2004).

Figure 8.1 illustrates the hypothetical example of a glucocorticoid thermal reaction norm graph for population B, showing the reaction norm intercept and the reaction norm slope. The XY plot of mean corrected integrated stress response versus environmental temperature can be plotted, and the slope of this XY plot determines the reaction norm. Figure 8.2 shows glucocorticoid thermal reaction norms for four subpopulations A, B, C, and D measured twice under good (stable) and extreme thermal regimes.

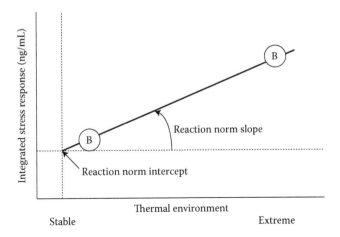

Figure 8.1 Hypothetical example of a glucocorticoid thermal reaction norm graph for population B, showing the reaction norm intercept and reaction norm slope. The XY plot of mean corrected integrated stress response versus environmental temperature can be plotted and the slope of this XY plot determines the reaction norm.

As shown in Figure 8.2, the shape of the glucocorticoid thermal reaction norm curve can be flat (population C) across temperatures. This suggests that the stress response of population C is independent of temperature. Typically, the variation in phenotypic traits decreases as distribution is more restricted, so one would expect this type of a reaction norm curve for animal species with a narrow geographical distribution (or species found at the edge of distribution limits), whereby it develops local adaptation characteristics to increase fitness under exposure to the given environmental conditions (Valladares et al. 2014). In the second scenario (Figure 8.2), populations B and D have glucocorticoid thermal reaction norm curves which are parallel, similar slopes. Therefore, there is likely to be no genetic variation in the plasticity and glucocorticoid responses of both populations that are equally dependent on the environment. Finally, populations B and A differ in their reaction norms (both in the intercept and slope) (Figure 8.2). Lendvai et al. (2014) have demonstrated the corticosterone reaction norms in house sparrows (*Passer domesticus*). Glucocorticoid thermal reaction norms of the two populations can be quantified and compared with other fitness traits (e.g., immunocompetence) to better understand their physiological sensitivity and adaptation under extreme environments.

Glucocorticoid thermal reaction norms provide a useful means of quantifying the physiological sensitivity of the stress endocrine responses in amphibians to acute shifts in environmental temperature toward extremes. Integration of quantitative measures of glucocorticoid thermal

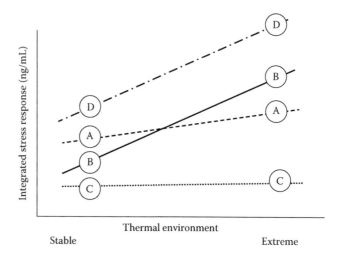

Figure 8.2 Glucocorticoid thermal reaction norms for four subpopulations A, B, C, and D measured under good (stable) and extreme thermal regimes. See text for explanations.

reaction norms with physiological and fitness-related data, such as reproductive rates, metabolic function, immune function, species abundance, and survival rates, could open up new ways of understanding whether plasticity in stress endocrine responses is beneficial or costly toward the animal's overall fitness (Chaput-Bardy et al. 2014; Lendvai et al. 2014). Studies should measure the glucocorticoid reaction norms of different individuals within a population under different environmental conditions. This will increase understanding of how selection shapes plasticity of liable hormone traits (Lendvai et al. 2014).

Conclusions

It is evident that global warming of the Earth's surface will eventually influence extreme events, such as heat waves. Therefore, extreme thermal events are important for defining the thermal limits of animal species, such as amphibians. Amphibians as ectothermic vertebrates regulate their intrinsic physiology and behaviors with respect to environmental temperature, and species living within fragile landscapes, such as mountain top endemics, might not have the physiological plasticity to cope with these episodic extreme temperatures. Therefore, using glucocorticoid thermal reaction norms, we can understand the scale of variation in the physical environment to which an organism responds, particularly how the stress endocrine response changes with environmental extremes, and what it means for the future adaptation and survival of species.

In conclusion, the thermodynamic response of corticosterone secretion in amphibians can provide insights into the sublethal effects of environmental temperature on amphibian-stress physiology. Details on the plasticity of corticosterone release could be obtained from glucocorticoid thermal reaction norms or even simply by assessing the variation in the individual responses to handling stress once acclimated to a temperature. This would give information on the phenotypic response (corticosterone release after a stressor) produced by the anuran genotype at different environmental conditions. It would also be worthwhile to use model anurans to investigate the impact of acute thermal changes using stress responses to mild handling stressors. Ultimately, the plasticity in corticosterone responses of amphibians with respect to thermal variability could be determined, expanding knowledge of how different amphibian species respond to and cope with increasing environmental temperatures. Furthermore, it has not been yet tested whether variability in temperature increases the severity of chytrid fungus infections. Hence, temperature variability could be the underlying link among climate change, disease risk, and amphibian declines. One way to address this significant gap will be via integrated, controlled studies on corticosterone stress responses of different species of frogs from different locations under different temperature gradients. Glucocorticoids, in particular corticosterone, have important effects on key physiological mechanisms, including immune function and metabolism. Hence, plasticity in corticosterone stress responses of different amphibian species will provide estimates of general responses to change and must be an important factor in the conservation and management of threatened amphibian species. Phenotypic plasticity in stress hormone responses under the influence of extreme environmental change could be important for animals for physiological resilience and persistence. Therefore, patterns of glucocorticoid thermal reaction norms will be important in quantifying species sensitivity under novel environments.

References

Alexander, M. A. and Eischeid, J. K. 2001. Climate variability in regions of amphibian declines. *Conserv. Biol.* 15: 930–942.

Bears, H., Smith, J. N. M., and Wingfield, J. C. 2003. Adrenocortical sensitivity to stress in dark-eyed Juncos (*Junco hyemalis oregonus*) breeding in low and high elevation habitat. *Ecoscience* 10: 127–133.

Beebee, T. J. C. 1995. Amphibian breeding and climate. *Nature* 374: 219–220.

Beebee, T. J. C. and Griffiths, R. A. 2005. The amphibian decline crisis: A watershed for conservation biology? *Biol. Conserv.* 125: 271–285.

Bern, H. and Nandi, J. 1964. Endocrinology of poikelothermic vertebrates. In Pincus, G., Thimann, K.V., and Astwood, F.B. (eds), *The Hormones*, Vol. IV. Academic Press, New York, pp. 199–293.

Blaustein, A. R., Belden, L. K., Olson, D. H., Green, D. M., Root, T. L., and Kiesecker. J. M. 2001. Amphibian breeding and climate change. *Conserv. Biol.* 15: 1804–1809.

Blieley, J. M. and Woodley, S. K. 2012. The effects of repeated handling and corticosterone treatment on behavior in an amphibian (Ocoee salamander: *Desmognathus ocoee*). *Physiol. Behav.* 105: 1132–1139.

Bly, J. E. and Clem, L. W. 1991. Temperature-mediated processes in teleost immunity: *In vitro* immunosuppression induced by *in vivo* low temperature in channel catfish. *Vet. Immun. Immunopathol.* 28: 365–377.

Bosch, J., Carrascal, L. M., Durán, L., Walker, S., and Fisher, M. C. 2007. Climate change and outbreaks of amphibian chytridiomycosis in a montane area of Central Spain; Is there a link? *Proc. R. Soc. B Biol. Sci.* 274: 253–260.

Bovo, R. P., Andrade, D. V., Toledo, L. F., Longo, A. V., Rodriguez, D., Haddad, C. F., Zamudio, K. R., and Becker, C. G. 2016. Physiological responses of Brazilian amphibians to an enzootic infection of the chytrid fungus Batrachochytrium dendrobatidis. *Dis. Aquat. Organ.* 117: 245–252.

Bozinovic, F., Calosi, P., and Spicer, J. I. 2011. Physiological correlates of geographic range in animals. *Annu. Rev. Ecol. Evol. Syst.* 42: 155–179.

Carey, C. and Alexander, M. A. 2003. Climate change and amphibian declines: Is there a link? *Divers. Distrib.* 9: 111–121.

Carey, C., Cohen, N., and Rollins-Smith, L. 1999. Amphibian declines: An immunological perspective. *Dev. Comp. Immunol.* 23: 459–472.

Carstensen, H., Burgers, A. C. J., and Li, C. H. 1961. Demonstration of aldosterone and corticosterone as the principal steroids formed in incubates of adrenals of the American bullfrog (*Rana catesbeiana*) and stimulation of their production by mammalian adrenocorticotropin. *Gen. Comp. Endocrinol.* 1: 37–50.

Chambers, D. L., Wojdak, J. M., Du, P., and Belden, L. K. 2013. Pond acidification may explain differences in corticosterone among salamander populations. *Physiol. Biochem. Zool.* 86: 224–232.

Chaput-Bardy, A., Ducatez, S., Legrand, D., and Baguette, M. 2014. Fitness costs of thermal reaction norms for wing melanisation in the large white butterfly (*Pieris brassicae*). *PLoS One* 9: e90026.

Clinchy, M., Sheriff, M. J., and Zanette, L. Y. 2013. Predator-induced stress and the ecology of fear. *Funct. Ecol.* 27: 56–65.

Corn, P. S. and Fogleman, J. C. 1984. Extinction of montane populations of the northern leopard frog (*Rana pipiens*) in Colorado. *J. Herpetol.* 18: 147–152.

Coumou, D. and Robinson, A. 2013. Historic and future increase in the global land area affected by monthly heat extremes. *Environ. Res. Lett.* 8: 034018 (034016pp).

Crump, M. L., Hensley, F. R., and Clark, K. L. 1992. Apparent decline of the golden toad: Underground or extinct? *Copeia* 1992: 413–420.

Dale, E. 1962. Steroid excretion by larval frogs. *Gen. Comp. Endocrinol.* 2: 171–176.

De Assis, V. R., Titon, S. C. M., Barsotti, A. M. G., Titon, B. R. Jr., and Gomes, F. R. 2015. Effects of acute restraint stress, prolonged captivity stress and transdermal corticosterone application on immunocompetence and plasma levels of corticosterone on the Cururu toad (*Rhinella icterica*). *PLoS One* 10: e0121005.

Denver, R. J. 2009. Stress hormones mediate environment-genotype interactions during amphibian development. *Gen. Comp. Endocrinol.* 164: 20–31.

Dickens, M. J., Delehanty, D. J., and Romero, M. L. 2009. Stress and translocation: Alterations in the stress physiology of translocated birds. *Proc. R. Soc. B* 276(1664): 2051–2056.

Dingemanse, N. J., Edelaar, P., and Kempenaers, B. 2010. Why is there variation in baseline glucocorticoid levels? *Trends Ecol. Evol.* 25: 261–262.

Dupont, W., Bourgeois, P., Reinberg, A., and Vaillant, R. 1979. Circannual and circadian rhythms in the concentration of corticosterone in the plasma of the edible frog (*Rana esculenta* L.). *J. Endocrinol.* 80: 117–125.

Florance, D., Webb, J. K., Dempster, T., Kearney, M. R., Worthing, A., and Letnic, M. 2011. Excluding access to invasion hubs can contain the spread of an invasive vertebrate. *Proc. R. Soc. B* 278: 2900–2908.

Francisco, B., Piero, C., and Spicer, J. I. 2011. Physiological correlates of geographic range in animals. *Annu. Rev. Ecol. Evol. Syst.* 42: 155–179.

Gabor, C. R., Fisher, M. C., and Bosch, J. 2013. A non-invasive stress assay shows that tadpole populations infected with *Batrachochytrium dendrobatidis* have elevated corticosterone levels. *PLoS One* 8: e56054.

Galloy, V. and Denoël, M. 2010. Detrimental effect of temperature increase on the fitness of an amphibian (*Lissotriton helveticus*). *Acta Oecol.* 36: 179–183.

Gervasi, S. S. and Foufopoulos, J. 2008. Costs of plasticity: Responses to desiccation decrease post-metamorphic immune function in a pond-breeding amphibian. *Funct. Ecol.* 22: 100–108.

Gibbs, J. P. and Breisch, A. R. 2001. Climate warming and calling phenology of frogs near Ithaca, New York, 1900–1999. *Conserv. Biol.* 15: 1175–1178.

Glennemeier, K. A. and Denver, R. J., 2002. Developmental changes in interrenal responsiveness in anuran amphibians. *Integr. Comp. Biol.* 42(3): 565–573.

Graham, C., Narayan, E., McCallum, H., and Hero, J.-M. 2013. Non-invasive monitoring of glucocorticoid physiology within highland and lowland populations of native Australian Great Barred Frog (*Mixophyes fasciolatus*). *Gen. Comp. Endocrinol.* 191: 24–30.

Hochachka, P. W. and Somero, G. N. 2002. *Biochemical Adaptation: Mechanism and Process in Physiological Evolution.* Oxford University Press, New York, 480p.

Hopkins, W. A. and DuRant, S. E., 2011. Innate immunity and stress physiology of eastern hellbenders (*Cryptobranchus alleganiensis*) from two stream reaches with differing habitat quality. *Gen. Comp. Endocrinol.* 174: 107–115.

Huey, R. B. and Kingsolver, J. G. 1989. Evolution of thermal sensitivity of ectotherm performance. *Trends Ecol. Evol.* 4: 131–135.

Hughes, L. 2003. Climate change and Australia: Trends, projections and impacts. *Austral. Ecol.* 28: 423–443.

Hume, D. M. and Egdahl, R. H. 1959. Effect of hypothermia and of cold exposure on adrenal cortical and medullary secretion. *Ann. N. Y. Acad. Sci.* 80: 435–444.

Ingram, G. J. 1990. The mystery of the disappearing frogs. *Wildlife Aust.* 27: 6–7.

IPCC (Intergovernmental Panel on Climate Change). 2007a. Contribution of working group I to the fourth assessment report of the Intergovernmental Panel on Climate Change. United Nations, Brussels.

IPCC. 2007b. Climate change 2007: Impacts, adaptations and vulnerability: Fourth assessment report of working group II. United Nations, Brussels.

Janin, A., Lena, J.-P., and Joly, P. 2011. Beyond occurrence: Body condition and stress hormone as integrative indicators of habitat availability and fragmentation in the common toad. *Biol. Conserv.* 144: 1008–1016.

Janin, A., Léna, J.-P., Deblois, S., and Joly, P. 2012. Use of stress-hormone levels and habitat selection to assess functional connectivity of a landscape for an amphibian. *Conserv. Biol.* 26: 923–931.

Jasnic, N., Djordjevic, J., Vujovic, P., Lakic, I., Djurasevic, S., and Cvijic, G. 2013. The effect of vasopressin 1b receptor (V1bR) blockade on HPA axis activity in rats exposed to acute heat stress. *J. Exp. Biol.* 216: 2302–2307.

Jessop, T. S., Letnic, M., Webb, J. K., and Dempster, T. 2013. Adrenocortical stress responses influence an invasive vertebrate's fitness in an extreme environment. *Proc. R. Soc. B* 280: 20131444.

Jungreis, A. M., Huibregtse, W. H., and Ungar, F. 1970. Corticosteroid identification and corticosterone concentration in serum of *Rana pipiens* during dehydration in winter and summer. *Comp. Biochem. Physiol.* 34: 683–689.

Jurani, M., Murgas, K., and Babusikova, F. 1973. Effect of stress and environmental temperature on adrenal function in *Rana esculenta*. *J. Endocrinol.* 57: 385–391.

Kindermann, C., Narayan, E., and Hero, J.-M. 2012. Urinary corticosterone metabolites and chytridiomycosis disease prevalence in a free-living population of male Stony Creek frogs (*Litoria wilcoxii*). *Comp. Biochem. Physiol. A Mol. Integr. Physiol.* 162: 171–176.

Laurance, W. F., McDonald, K. R., and Speare, R. 1996. Epidemic disease and the catastrophic decline of Australian rainforest frogs. *Conserv. Biol.* 10: 406–413.

Laurance, W. F., Carolina, U. D., Shoo, L. P., Herzog, S. K., Kessler, M., Escobar, F., Brehm, G., Axmacher, J. C., Chen, I.-C., and Gámez, L. A. 2011. Global warming, elevational ranges and the vulnerability of tropical biota. *Biol. Conserv.* 144(1): 548–557.

Leboulenger, F., Delarue, C., Tonon, M. C., Jegou, S., and Vaudry, H. 1978. *In vitro* study of frog (Rana ridibunda Pallas) interrenal function by use of a simplified perifusion system: I. Influence of adrenocorticotropin upon corticosterone release. *Gen. Comp. Endocrinol.* 36: 327–338.

Leboulenger, F., Delarue, C., Belanger, A., Perroteau, I., Netchitailo, P., Leroux, P., Jegou, S., Tonon, M. C., and Vaudry, H. 1982. Direct radioimmunoassays for plasma corticosterone and aldosterone in frog. I. Validation of the methods and evidence for daily rhythms in a natural environment. *Gen. Comp. Endocrinol.* 46: 521–532.

Leboulenger, F., Trochard, M. C., Morin, J. P., Dupont, W., Vaudry, H., and Vaillant, R. 1977. Competitive protein-binding radioassay of corticosterone in the plasma of the frog, *Rana esculenta* L. (author's transl). *J. Physiol.* 73: 73–83.

Lendvai, Á. Z., Ouyang, J. Q., Schoenle, L. A., Fasanello, V., Haussmann, M. F. et al. 2014. Experimental food restriction reveals individual differences in corticosterone reaction norms with no oxidative costs. *PLoS One* 9(11): e110564.

Licht, P., McCreery, B. R., Barnes, R., and Pang, R. 1983. Seasonal and stress related changes in plasma gonadotropins, sex steroids, and corticosterone in the bullfrog, *Rana catesbeiana*. *Gen. Comp. Endocrinol.* 50: 124–145.

Lofts, B., Wellen, J. J., and Benraad, T. S. 1972. Seasonal changes in endocrine organs of the male common frog, *Rana temporaria* III. The gonads and cholesterol cycle. *Gen. Comp. Endocrinol.* 18: 344–363.

Maniero, G. D. and Carey, C. 1997. Changes in selected aspects of immune function in the leopard frog, *Rana pipiens*, associated with exposure to cold. *J. Comp. Physiol. B.* 167: 256–263.

Matutte, B., Storey, K. B., Knoop, F. C., and Conlon, J. M. 2000. Induction of synthesis of an antimicrobial peptide in the skin of the freeze-tolerant frog, *Rana sylvatica*, in response to environmental stimuli. *FEBS Lett.* 483: 135–138.

McEwen, B. S. and Stellar, E. 1993. Stress and the individual. *Arch. Intern. Med.* 153: 2093–2101.

McEwen, B. S. and Wingfield, J. C. 2007. Allostasis and allostatic load. In Fink, G. (ed.), *Encyclopedia of Stress*, 2nd edn. Academic Press, New York, pp. 135–141.

McNab, B. K. 2002. *The Physiological Ecology of Vertebrates: A View from Energetics.* Comstock Publishing Associates, Cornell University Press, Ithaca, NY.

Monfort, S. L. 2003. Non-invasive endocrine measures of reproduction and stress in wild populations. In Holt, W. V., Pickard, A. R., Rodger, J. C., and Wildt, D. E. (eds), *Reproductive Science and Integrated Conservation.* Cambridge University Press, Cambridge, pp. 147–165.

Morley, S. A., Martin, S. M., Day, R. W., Ericson, J., Lai, C.-H. et al. 2012. Thermal reaction norms and the scale of temperature variation: Latitudinal vulnerability of intertidal nacellid limpets to climate change. *PLoS One* 7: e52818.

Nandi, J. 1967. Comparative endocrinology of steroid hormones in vertebrates. *Am. Zool.* 7: 115–133.

Narayan, E. J., Cockrem, J. F., and Hero, J.-M. 2011. Urinary corticosterone metabolite responses to capture and captivity in the cane toad (*Rhinella marina*). *Gen. Comp. Endocrinol.* 173: 371–377.

Narayan, E., Cockrem, J. F., and Hero, J.-M. 2012. Effects of temperature on urinary corticosterone metabolite responses to short-term capture and handling stress in the cane toad (*Rhinella marina*). *Gen. Comp. Endocrinol.* 78: 301–305.

Narayan, E. and Hero, J.-M. 2014. Acute thermal stressor increases glucocorticoid response but minimizes testosterone and locomotor performance in the cane toad (*Rhinella marina*). *PLoS One* 9: e92090.

Narayan, E., Molinia, F., Christi, K., Morley, C., and Cockrem, J. 2010. Urinary corticosterone metabolite responses to capture, and annual patterns of urinary corticosterone in wild and captive endangered Fijian ground frogs (*Platymantis vitiana*). *Aust. J. Zool.* 58: 189–197.

Nguyen, K. D. T., Morley, S. A., Lai, C.-H., Clark, M. S., Tan, K. S. et al. 2011. Upper temperature limits of tropical marine ectotherms: Global warming implications. *PLoS One* 6: e29340.

Nyström, P., Hansson, J., Månsson, J., Sundstedt, M., Reslow, C., and Broström, A. 2007. A documented amphibian decline over 40 years: Possible causes and implications for species recovery. *Biol. Conserv.* 138: 399–411.

Ouyang, J. Q., Hau, M., and Bonier, F. 2011. Within seasons and among years: When are corticosterone levels repeatable? *Horm. Behav.* 60: 559–564.

Pancak, M. K. and Taylor, D. H. 1983. Seasonal and daily plasma corticosterone rhythms in American toads, *Bufo americanus. Gen. Comp. Endocrinol.* 50: 490–497.

Parmesan, C. 2006. Ecological and evolutionary responses to recent climate change. *Annu. Rev. Ecol. Syst.* 37: 637–669.

Pearson, R. G. and T. P. Dawson. 2005. Long-distance plant dispersal and habitat fragmentation: Identifying conservation targets for spatial landscape planning under climate change. *Biol. Conserv.* 123: 389–401.

Phillips, B. L. and Puschendorf, R. 2013. Do pathogens become more virulent as they spread? Evidence from the amphibian declines in Central America. *Proc. Biol. Sci.* 7: 2013290.

Pounds, J. A., Fogden, M. P. L., and Campbell, J. H. 1999. Biological response to climate change on a tropical mountain. *Nature* 398: 611–615.

Pounds, J. A., Fogden, M. P. L., Savage, J. M., and Gorman, G. C. 1997. Tests of null models for amphibian declines on a tropical mountain. *Conserv. Biol.* 11: 1307–1322.

Preest, M. R., Cree, A., and Tyrrell, C. L. 2005. ACTH-induced stress response during pregnancy in a viviparous gecko: No observed effect on offspring quality. *J. Exp. Zool.* 303: 823–835.

Prosser, C. L. 1973. *Comparative Animal Physiology.* Saunders, Philadelphia, PA.

Pxytycz, B. and Jozkowicz, A. 1994. Differential effects of temperature on macrophages of ectothermic vertebrates. *J. Leukoc. Biol.* 56: 729–731.

Raffel, T. R., Rohr, J. R., Kiesecker, J. M., and Hudson, P. J. 2006. Negative effects of changing temperature on amphibian immunity under field conditions. *Funct. Ecol.* 20: 819–828.

Raffel, T. R., Romansic, J. M., Halstead, N. T., McMahon, T. A., Venesky, M. D., and Rohr, J. R. 2012. Disease and thermal acclimation in a more variable and unpredictable climate. *Nat. Clim. Change* 3: 146–151.

Reading, C. 2007. Linking global warming to amphibian declines through its effects on female body condition and survivorship. *Oecologia* 151: 125–131.

Rich, E. L. and Romero, L. M. 2005. Exposure to chronic stress downregulates corticosterone responses to acute stressors. *Am. J. Physiol. Regul. Integr. Comp. Physiol.* 288: R1628–R1636.

Rollins-Smith, L. A. 2001. Neuroendocrine-immune system interactions in amphibians: Implications for understanding global amphibian declines. *Immunol. Res.* 23: 273–280.

Romero, L. M. 2004. Physiological stress in ecology: Lessons from biomedical research. *Trends Ecol. Evol.* 19: 249–255.

Root, T. L., Price, J. T., Hall, K. R., Schneider, S. H., Rosenzweig, C., and Pounds, J. L. 2003. Finger-prints of global warming on wild animals and plants. *Nature* 421: 37–42.

Sterling, P. and Eyer, J. 1988. Allostasis: A new paradigm to explain arousal pathology. In Fisher, S. and Reason, J. (eds), *Handbook of Life Stress, Cognition and Health.* John Wiley, New York, pp. 629–647.

Stewart, M. M. 1995. Climate driven population fluctuations in rain-forest frogs. *J. Herpetol.* 29: 437–446.

Stuart, S., Chanson, J. S., Cox, N. A., Young, B. E., Rodrigues, A. S. L., Fishman, D. L., and Waller, R. W. 2004. Status and trends of amphibian declines and extinctions worldwide. *Science* 306(5702): 1783–1786.

Tingley, R., Greenlees, M. J., and Shine, R. 2012. Hydric balance and locomotor performance of an anuran (*Rhinella marina*) invading the Australian arid zone. *Oikos* 121: 1959–1965.

Valladares, F., Matesanz, S., Guilhaumon, F., Araujo, M. B., Balaguer, L. et al. 2014. The effects of phenotypic plasticity and local adaptation on forecasts of species range shifts under climate change. *Ecol. Lett.* 17: 1351–1364.

Wack, C. L., DuRant, S. E., Hopkins, W. A., Lovern, M. B., Feldhoff, R. C., and Woodley, S. K. 2012. Elevated plasma corticosterone increases metabolic rate in a terrestrial salamander. *Comp. Biochem. Physiol. A Mol. Biochem. Integr. Physiol.* 161: 153–158.

Wake, D. B. and Vredenburg, V. T. 2008. Are we in the midst of the sixth mass extinction? A view from the world of amphibians. *PNAS* 105: 11466–11473.

Wasser, S. K., Hunt, K. E., Brown, J. L., Cooper, K., Crockett, C. M., Bechert, U., Millspaugh, J. J., Larson, S., and Monfort, S. L. 2000. A generalized fecal glucocorticoid assay for use in a diverse array of nondomestic mammalian and avian species. *Gen. Comp. Endocrinol.* 120: 260–275.

Woodhams, D. C., Alford, R. A., Briggs, C. J., Johnson, M., and Rollins-Smith, L. A. 2008. Life-history trade-offs influence disease in changing climates: Strategies of an amphibian pathogen. *Ecology* 89: 1627–1639.

Index